Lockout/Tagout:

A Practical Approach

Stephen M. Kelley, CHMM, CSP

American Society of Safety Engineers ❖❖ Des Plaines, Illinois USA

Lockout/Tagout: A Practical Approach

Library of Congress Cataloging-in-Publication Data

Kelley, Stephen M.
Lockout/tagout: a practical approach / Stephen M. Kelley
p. cm
Includes index.
ISBN 1-885581-35-1 (alk. paper)
1. Industrial safety—Handbooks, manuals, etc. 2. Power (Mechanics)—Handbooks, manuals, etc. 3. Energy transfer—Handbooks, manuals, etc. I. Title.
T55 .K445 2001
658.2'02—dc21 2001027943

American Society of Safety Engineers
1800 East Oakton Street
Des Plaines, Illinois 60018-2187
USA

This publication is designed to provide accurate and authoritative information in regard to the subject matter covered. It is provided with the understanding that the author is not hereby engaged in rendering legal or other professional services. If legal advice or other professional assistance is required, the services of a competent professional should be sought. While this book was carefully produced, the author and publisher do not warrant the information contained herein to be free from errors.

Project Editor: Charles T. Coffin
Copy Editor: Nancy E. Kaminski
Text Design, Composition and Layout: Nancy E. Kaminski
Cover Design: Publication Design, Inc.

Printed in the United States on acid-free paper

3 4 5 6 7 8

Table of Contents

Preface

Lockout/Tagout—A Practical Approach is a basic handbook for achieving compliance with OSHA's Lockout/Tagout standard. This book is intended to relate practical considerations and strategies for implementing a comprehensive energy control program. Potential accident situations are examined, and case studies are presented to highlight common causes of energy-related accidents and effective practices for ensuring full protection of personnel during the service and maintenance of machines, equipment and processes. Energy isolation techniques, protective appliances, and verification methods for specific energy types are described, including special considerations necessary when employees work on high-voltage circuits and equipment. Complex lockout/tagout situations, including group lockout/tagout procedures, and procedures for ensuring the continuity and orderly transfer of lockout/tagout protection during shift or personnel changes are also discussed.

In this era of tight budgets and lean staffing, responsibility for administering lockout/tagout and other safety programs often falls to Human Resources, Production, or Facilities managers with little or no formal experience in safety administration. With this in mind, critical success factors for an effective lockout/tagout program are outlined, and advice is provided on avoiding common pitfalls. Various graphic illustrations, checklists, and forms, along with an example written lockout/tagout program and a sample format for equipment-specific energy control procedures, are provided to assist safety practitioners at all levels in launching, maintaining or improving their lockout/tagout program.

Acknowledgments

My first thanks go to my wife Lori, and children Catherine and Elizabeth; their endless support and patience made completion of this book possible. For their confidence and encouragement, I am grateful to my parents, and to Frankie, Tricia, Ellen, Lynda, Julie, Matt and Peggy. My admiration and thanks also go to Ed Eagan, Bob Walsh, Mike Diggin, Jim Manoli, Kevin Biernacki, and Hal Basdekis, whose competence and commitment to worker safety have undoubtedly prevented countless accidents and injuries. Special thanks also go to Professor Carol L. Russell of Framingham State College for her sage advice and counsel. Finally, I extend my gratitude to the Boston Latin School, the Liberty Mutual Research Center for Safety and Health, and the American Society of Safety Engineers for their positive influences and continuing commitment to excellence.

Chapter 1

Introduction

Continuous advances in machine, equipment, and process technology are making companies more efficient, productive, and competitive—necessary attributes in a global economy. As technology has advanced, however, the energy sources driving machines have become more powerful and sophisticated, making service and maintenance tasks more demanding, complex—and often—more dangerous. While machine safeguarding has also progressed, injury risks persist for those who operate, service, and maintain equipment.

OSHA's final rule on Control of Hazardous Energy (Lockout/Tagout) was issued on September 1, 1989. The standard requires that machines and equipment be de-energized and rendered inoperative before performing any service or maintenance, where unexpected startup or release of hazardous energy could occur and cause injury. Studies which showed the failure to control hazardous energy to be a widespread problem provided the basis for the lockout/tagout rulemaking. For example, the OSHA study *Selected Occupational Fatalities Related to Lockout/Tagout Problems as Found in Reports of OSHA Fatality/Catastrophe Investigations* identified 83 fatalities from 1974 to 1980 that were caused by failure to properly de-energize machines prior to performing maintenance, repairs, or servicing. Also, the National Institute for Occupational Safety and Health's (NIOSH) *Study of Hazardous Release of Energy Injuries in Ohio in 1983* documented 339 accidents that resulted from unexpected release of energy during equipment repair, servicing, or maintenance in Ohio alone during 1983.

The lockout/tagout standard was expected to prevent approximately 122 fatalities, 28,400 lost workday injuries, and 31,900 non-lost workday injuries annually, yet accidents associated with the release of hazardous energy continue to occur. Although exact data on the number of workers killed or injured as a result of contact with uncontrolled hazardous energy are unavailable, several recent studies highlight the continuing extent of the problem. During the period from 1982–1997, NIOSH investigated 1,281 fatalities as part of their Fatality Assessment and Control Evaluation (FACE) Program. Of the 1,281 total, 152 deaths involved installation or maintenance, service, or repair tasks on or near machines, equipment, processes, or systems. Since the FACE program included only 20 states, the fatalities investigated represent only a portion of the total deaths caused by contact with uncontrolled hazardous energy. A 1998 Bureau of Labor Statistics (BLS) study (*Worker Fatalities from being Caught in Machinery)* reported an average of almost 150 deaths and 1000 amputations a year from workers being caught in machinery in the period from 1992–96. A six year high of 189 workplace deaths were attributed to the same cause in 1997, a 29 percent increase over 1996 (146 workplace deaths). More than half of these fatalities occurred while workers were carrying out service- or maintenance-related tasks. It is these very types of accidents that the OSHA lockout/tagout standard is meant to prevent.

The 1998 BLS study reveals that myriad industries and occupations remain exposed to hazardous energy risks. Of the 189 fatalities caused by workers being caught in equipment in 1997, most occurred in the manufacturing sector (41%), followed by agriculture, forestry, and fishing (26%), construction (9%), mining (6%), transportation and public utilities (6%), wholesale trade (6%), and services (4%). The machine-related fatalities and injuries that have occurred cut across all industry sectors.

The study also revealed the varied nature of the tasks and equipment types involved in machine-related fatalities. Material handling machinery such as conveyors, cranes, and elevators accounted for 18% of reported fatalities, the same as construction, logging, and mining machinery (18%), followed by metal, woodworking, and special machinery (17%), agricultural and garden machinery (15%), special process and packaging machinery (14%), vehicles (9%), and miscellaneous machinery (5%). A wide range of activities was also being performed by workers who became fatally caught in equipment. For example 45% of the fatalities reported in the study occurred while workers were operating equipment, 20% while performing general repair or maintenance, 10% while unjamming materials, 7% while cleaning machines, 4% while adjusting machines, 4% while reaching or stepping over machines, and 3% while inspecting or checking machines. Tables 1.1 through 1.3 tabulate the results of the study.

Work-related electrocutions also provide an indication of the scope of the problem. According to the BLS' *National Census of Fatal Occupational Injuries,* electrocutions accounted for 6 percent (361) of the 6,026 fatal work injuries that occurred in 1998, an increase of 12% over 1997 (318). In 1996, 279 electrocutions were reported, accounting for 5 percent of worker deaths. Electricians and their apprentices accounted for about one-fifth of the electrocutions, and mechanics and repairers accounted for one-tenth. While contact with overhead power lines accounted for the highest number of deaths from electrocution, many workers were electrocuted while installing or repairing machines, appliances, and other equipment.

In 1998, lockout/tagout violations ranked fourth on OSHA's top ten list of most frequently cited standards (3,532 violations), and fourth on the top ten list of serious violations (2,537 serious violations). Despite the alarming number of machine-related accidents and OSHA's enforcement efforts, lockout/tagout is still viewed by many as a burdensome, production-slowing task. Companies that fail to recognize and control machine hazards face regulatory fines, and risk the occurrence of a serious and costly machine-related accident. With an effective lockout/tagout program, however, injuries and losses arising from failure to control hazardous energy are entirely preventable.

Table 1.1: Worker fatalities from being caught in machinery, by selected industry, 1997

Industry	Number	Percent
Agriculture, forestry, and fishing	50	26
Agricultural production, crops	30	16
Agricultural production, livestock	12	6
Agricultural services	5	3
Mining	11	6
Construction	17	9
Heavy construction, except building	10	5
Special trade contractors	6	3
Manufacturing	78	41
Food and kindred products	11	6
Lumber and wood products	11	6
Paper and allied products	8	4
Chemicals and allied products	4	2
Rubber and miscellaneous plastics products	5	3
Stone, clay, and glass products	9	5
Primary metal industries	9	5
Fabricated metal products	8	4
Industrial machinery and equipment	7	4
Transportation and public utilities	11	6
Trucking and warehousing	3	2
Electric, gas, and sanitary services	5	3
Refuse systems	5	3
Wholesale trade	11	6
Wholesale trade, durable goods	8	4
Scrap and waste materials	7	4
Wholesale trade, nondurable goods	3	2
Services	7	4
TOTAL	189	100

Source: US Department of Labor, Bureau of Labor Statistics, *Worker Fatalities from being Caught in Machinery.* (Washington DC: GPO, 1998).

Table 1.2: Selected machinery involved in fatalities resulting from workers being caught in machinery, 1997

Type of machinery involved	Number	Percent
Agriculture and garden machinery	28	15
Harvesting and threshing machinery	15	8
Balers	8	4
Harvesters, reapers	4	2
Mowing machinery	4	2
Construction, logging, and mining machinery	34	18
Excavating machinery	4	2
Loaders	4	2
Logging and wood processing machinery–specialized	7	4
Debarkers	3	2
Mining and drilling machinery	13	7
Drilling machines, drilling augers	7	4
Other construction, logging, and mining machinery	6	3
Agitators, mixers–earth, mineral	3	2
Compactors, crushers, pulverizers–earth, mineral	3	2
Material handling machinery	34	18
Conveyors–powered	23	12
Conveyors–belt	10	5
Conveyors–live roller	3	2
Conveyors–screw, auger	4	2
Cranes	5	3
Elevators	3	2
Metal, woodworking, and special material machinery	32	17
Bending, rolling, and shaping machinery	3	2
Extruding, injecting, forming, molding machinery	11	6
Casting machinery	5	3
Plastic injection molding machinery	4	2
Lathes	3	2

Table 1.2: Selected machinery involved in fatalities resulting from workers being caught in machinery, 1997 (Continued)

Type of machinery involved	Number	Percent
Presses, except printing	4	2
Sawing machinery–stationary	6	3
Special process machinery	26	14
Food and beverage processing machinery–specialized	4	2
Packaging, bottling, wrapping machinery	7	4
Paper production machinery	4	2
Textile, apparel, leather production machinery	3	2
Other special process machinery	7	4
Pumps	4	2
Vehicles	18	9
Truck	6	3
Forklift	6	3
Tractor	5	3
Miscellaneous machinery	9	5
Trash compactors	5	3
Electronic doors, gates	3	2
TOTAL	189	100

Source: US Department of Labor, Bureau of Labor Statistics, *Worker Fatalities from being Caught in Machinery.* (Washington DC: GPO. 1998).

Table 1.3: Activity performed by workers fatally caught in machinery, 1997

Industry	Number	Percent
Operating machine	85	45
General (unspecified) repair or maintenance	37	20
Unjamming materials	19	10
Cleaning machine	14	7
Adjusting machine	8	4
Reaching or stepping over machine	7	4
Inspecting or checking machine	5	3
Other	14	7
TOTAL	189	100

Source: US Department of Labor, Bureau of Labor Statistics, *Worker Fatalities from being Caught in Machinery.* (Washington DC: GPO. 1998).

Case Studies

Accidents involving contact with uncontrolled hazardous energy (improper lockout) often result in serious injury or death. OSHA estimates that failure to adequately control hazardous energy accounts for *7% of all occupational fatalities and nearly 10 percent of serious accidents* in many industries. The following case studies are examples of serious accidents that provide insight into how hazardous energy releases can occur unexpectedly.

Case 1

A 22-year-old laborer was steam cleaning a scraper. The bowl apron had been left in the raised position. The hydraulically controlled apron had not been lowered or blocked to prevent it from accidentally falling. The apron fell unexpectedly and the employee was caught between the apron and the cutting edge of the scraper bowl and killed. The apron weighed approximately 2500 pounds.

Case 2

A 53-year-old journeyman electrician was attempting to correct an electrical problem involving two non-operational lamps. He proceeded to the area where he thought the problem was. He had not shut off the power at the circuit breaker panel, nor had he tested the wires to see if they were live. He was electrocuted when he grabbed the two live wires with his left hand and then fell from the ladder.

Case 3

A 38-year-old worker at a county sanitary landfill died after falling into a large trash compactor used to bale cardboard for recycling. The cardboard was lifted 20 feet by a belt conveyor and fed through a 20- by 44-inch opening into a hopper. The hopper had automatic controls that activated the baler when enough material collected in the baling chamber. When the baler was activated, material in the chamber was compressed by a ram that entered the chamber from the side. Excess material above the chamber was trimmed by a shearer.

On the day of the incident, cardboard jammed at the conveyor discharge opening. Without stopping, de-energizing, or locking out the equipment, the victim rode the conveyor up to the discharge opening to clear the jam. He fell into the hopper and the baling cycle was automatically activated, amputating his legs. The victim bled to death before he could be removed from the machine [Colorado Department of Public Health and Environment 1994].

Case 4

A 53-year-old journeyman wireman was electrocuted when he contacted two energized, 6.9-kilovolt buss terminals. The victim and two coworkers (all contract employees) were installing electrical components of a sulfur dioxide emission control system in a 14-compartment switch house. The circuit breaker protecting the internal buss within the switch house had been tripped out and marked with a tag—but it had not been secured by locking. This procedure was consistent with the hazardous energy control procedures of the power plant.

The victim and his coworkers were wiping down the individual compartments before a pre-startup inspection by power plant personnel. Without the knowledge of the victim and his coworkers, power plant personnel had energized the internal buss in the switch house. When the victim began to wipe down one of the compartments at the south end of the switch house, he contacted the A-phase buss terminal with his right hand and the C-phase buss terminal with his left hand. This act completed a path between phases, and the victim was electrocuted.

A coworker walking past the victim during the incident was blown backward by the arc blast and received first-degree flash burns on his face and neck. A second coworker at the north end of the switch house heard the explosion and came to help. He notified the contractor's safety coordinator by radio and requested EMS. The EMS responded in about 15 minutes and transported the victim to a local hospital emergency room where he was pronounced dead [NIOSH 1994].

Case 5

A 33-year-old janitorial worker died after he was trapped inside a linen dryer at a hospital laundry while cleaning plastic debris from the inside of the dryer drum. The cleaning task (which usually took 15 minutes to an hour) involved propping open the door to the dryer with a piece of wood and entering the 4- by 8-foot dryer drum. The melted debris was removed by scraping and chiseling it with screwdrivers and chisels. The dryer was part of an automated system that delivered wet laundry from the washer through an overhead conveyor to the dryer, where it was dried during a 6-minute cycle with air temperatures of 217° to 230°F. The system control panel was equipped with an error light that was activated if the dryer door was open, indicating that the dryer was out of service.

On the night of the incident, the victim propped the door open and entered the dryer drum without de-energizing or locking out the dryer. He began to clean the inside of the drum. Although the error light had been activated when the door was propped open, the signal was misinterpreted by a coworker, who restarted the system. When the system was restarted, the overhead conveyor delivered a 200-pound load of wet laundry to the dryer, knocking out the wooden door prop, trapping the victim inside, and automatically starting the drying cycle. The victim remained trapped inside until the cycle was completed and was discovered when the load was discharged from the dryer. He died thirty minutes later of severe burns and blunt head trauma [Massachusetts Department of Public Health 1992].

Causal Factors

A study by OSHA's Office of Data Analysis entitled *Occupational Fatalities Related to Fixed Machinery as Found in Reports of OSHA Fatality/Catastrophe Investigations* analyzed 125 fatalities involving fixed machinery that occurred between 1974 and 1976. The causes of these fatalities cited by the study include:

- failure to adhere to safe operating procedures;
- accidental machine activation;
- machine not deactivated; and
- equipment failure.

The continued occurrence of energy-related accidents prompted NIOSH to issue an Alert entitled *Preventing Worker Deaths from Uncontrolled Release of Electrical, Mechanical, and Other Types of Hazardous Energy* in August 1999. The Alert identified three related factors which contribute to injuries and deaths:

- failure to completely de-energize, isolate, block, and dissipate the hazardous energy source;
- failure to lockout *and* tagout energy control devices and isolation points after the hazardous energy source has been de-energized; and
- failure to verify that the hazardous energy source was de-energized before performing work.

While the exact causes of accidents involving improper energy control vary, they can generally be attributed to management system failures. Management errors such as inadequate training, poor energy control procedures, and a lack of lockout/tagout program emphasis all increase the likelihood of accidents. What is clear from the body of research on energy-related accidents is that they are preventable if effective energy control procedures are developed and used.

This book is intended to provide the reader with knowledge and skill for preventing energy-related accidents. In the chapters that follow, an effective approach for developing a lockout/tagout program is outlined, and key considerations for success are discussed. Management factors that contribute to the success or failure of energy control programs are also examined. Chapter 3, *Methods of Energy Control,* describes the different types of hazardous energy and specific techniques for energy control.

Subsequent chapters describe the "how to" of lockout/tagout, including preparation for lockout, sequence and application of lockout, verification of de-energization, and release from lockout. Specific chapters are also dedicated to discussing group lockout/tagout and transfer of lockout/tagout protection, with examples of industry practices provided. Chapters are also dedicated to vehicles and equipment, and employee training. A chapter on *Energy Control Procedures* (ECP) outlines when ECPs need to be written and what is required in an ECP. A sample ECP format is provided.

Other chapters discuss requirements and provide guidelines for complying with other mandated lockout/tagout program elements, including protective materials and hardware, installation and design, and periodic inspections. The final chapter, *Beyond OSHA Compliance,* describes tips and techniques for ensuring lockout/tagout program success, and for going beyond OSHA's minimum requirements to ensure full employee protection during service and maintenance of machines.

Chapter 2

Getting Started

Like any other project, developing and implementing a lockout/tagout program first requires planning. Specific tasks must be identified, responsibilities must be assigned, and standards of performance established to direct and sustain the program. An individual (or individuals) will be needed to champion the program, set the plan into motion, and motivate employees. Most often, the company safety representative will be called upon to spearhead the effort. In larger companies with numerous or complex machines and equipment, work teams or task groups will need to be formed and empowered. Consultants can also help companies with program development, implementation, and training, and can be of great assistance to companies that lack necessary in-house resources.

Scope and Applicability of the Lockout/Tagout Standard

A goal of every safety program, regardless of industry, should be to protect employees from machine, equipment and process hazards. In addition, it is important from a compliance perspective to be familiar with the particular standards applicable to your operations. The lockout/tagout standard set forth in 1910.147 covers the servicing and maintenance of machines, equipment, and processes (MEP) in *general industry.* The scope of the standard does not cover:

- construction, agriculture, and maritime employment;
- installations under the exclusive control of electric utilities for the purpose of power generation, transmission, and distribution, including related equipment for communication or metering;
- exposure to electrical hazards from work on, near, or with conductors or equipment in electric utilization installations (these are covered by Subpart S of Part 1910); and
- oil and gas well drilling and servicing.

OSHA exempted the above industries and limited the scope of the standard to general industry because their unique situations and work practices would have unduly complicated the standard. OSHA may, however, develop requirements for the control of hazardous energy in these industries as part of future rule-making proceedings.

In addition to the lockout/tagout standard, a number of other OSHA general standards impose lockout-related requirements and are listed in Table 2.1. These standards prescribe various design requirements for certain machines, equipment, or industries, but do not address lockout/tagout procedures in any detail. The procedural and training requirements of 1910.147 supplement the energy control provisions of these other standards.

Table 2.1: Other OSHA standards with lockout/tagout-related requirements

Standard	Number	Sections
Permit-required Confined Spaces	1910.146	(d)(3)
Servicing Multi-piece and Single-piece Rim Wheels	1910.177	(c)(2)(iv), (d)(1–3)
Powered Industrial Trucks	1910.178	(q)(5)(i)
Overhead and Gantry Cranes	1910.179	(g)(5)(i–iii), (l)(2)(i)(b–d)
Derricks	1910.181	
Woodworking Machinery	1910.213	(a)(10), (b)(5)
Mechanical Power Presses	1910.217	(b)(8)(i), (d)(9)(iv)
Forging Machines	1910.218	(a)(3)(iii, iv), (d)(2), (e)(1)(ii, iii), (f)(1)(i–iii), (f)(2)(i, ii), (g)(2), (h)(2, 5), (i)(1, 2), (j)(1), (h)(5)
Pulp, Paper, and Paperboard Mills	1910.261	(b)(4), (f)(6)(i), (g)(2), (g)(15)(i), (g)(19)(iii), (j)(4)(iii), (j)(5)(iii), (k)(2)(ii),
Textiles	1910.262	(c)(1), (n)(2), (q)(2), (p)(1)
Bakery Equipment	1910.263	(l)(3)(iii)(b), (l)(8)(iii)
Sawmills	1910.265	(c)(13), (c)(26)(v)
Electric Power Generation, Transmission, and Distribution	1910.269	Miscellaneous
Grain Handling	1910.272	(e)(1)(ii), (g)(1)(ii), (l)(4)
Wiring Methods, Components, and Equipment for General Use	1910.305	(j)(4)(ii)(A), (j)(4)(ii)(c)(1)
Specific Purpose Equipment and Installations	1910.306	Miscellaneous
Selection and Use of Work Practices	1910.333	Miscellaneous

OSHA's construction industry standards also include various lockout/tagout-related provisions for deactivating equipment, de-energizing and grounding electrical circuits, and limiting employee access to energized parts in construction work. These include:

- General Requirements 1926.416
- Lockout and Tagging of Circuits 1926.417

Similarly, OSHA's standards for maritime employment include various provisions for de-energizing equipment on vessels and in marine terminals while undergoing service or maintenance. These include:

- Ship's Boilers 1915.162
- Ship's Piping Systems 1915.163
- Ship's Propulsion Machinery............... 1915.164
- Ship's Deck Machinery.................... 1916.165
- Electrical Circuits and Distribution Boards ... 1915.181
- Conveyors .... 1917.48(i)
- Machine Guarding .... 1917.151(b)

With respect to applicability, the lockout/tagout standard generally applies to the control of hazardous energy during the servicing and maintenance of machines, equipment, and processes.

> 1910.147 (a)(1)(i)
>
> This standard covers the servicing and maintenance of machines and equipment in which the unexpected energization or start up of the machines or equipment, or release of stored energy could cause injury to employees. This standard establishes minimum performance requirements for the control of such hazardous energy.

According to OSHA's Lockout/Tagout Instruction Standard, 29 CFR 1910.147 also applies to high-intensity electromagnetic fields regulated at 29 CFR 1910.97—*Nonionizing Radiation*. Consequently, electromagnetic devices must be de-energized and held off whenever workers are present within a high-intensity ambient field.

Specific activities and operations, however, are exempted from coverage, including:

- Work on cord and plug connected electric equipment for which the exposure to the hazards of unexpected energization or start up of the equipment is controlled by unplugging the equipment from the energy source and by the plug being under the exclusive control of the employee performing the servicing and maintenance; and
- Hot tap operations involving transmission and distribution systems for substances such as gas, steam, water, or petroleum products when they are performed on pressurized pipelines,

provided that the employer demonstrates that: (1) continuity of service is essential; (2) shutdown of the system is impractical; and (3) documented procedures are followed, and special equipment is used which will provide proven effective protection for employees.

- Servicing/maintenance of fire alarm and extinguishing systems and their components, upon which other employees are dependent for fire safety, if the workers performing the servicing or maintenance are protected from hazards related to the unexpected release of hazardous energy by appropriate alternative measures.

In determining which other activities *are* covered by the lockout/tagout standard, one must understand the difference between "servicing and maintenance," and "normal production operations." OSHA defines service and maintenance in the standard as "[w]orkplace activities such as constructing, installing, setting up, adjusting, inspecting, modifying, and maintaining and/or servicing machines or equipment." These types of activities can expose workers to the unexpected energization or start up of machines and equipment, or to the release of hazardous energy, and lockout/tagout must be used pursuant to 1910.147. OSHA believes that the vast majority of service and maintenance tasks can only be performed with equipment locked and tagged out.

In contrast, normal production operations, defined as "[t]he utilization of a machine or equipment to perform its intended production function," are generally *not* covered by the standard. An exception noted in the lockout/tagout standard in paragraph (a)(2)(ii) defines the types of activities associated with normal production operations, and that would therefore not be covered by the standard. The exception states:

> 1910.147(a)(2)(ii)
>
> Minor tool changes and adjustments, and other minor servicing activities, which take place during normal production operations are not covered by the standard if they are routine, repetitive, and integral to the use of the equipment for production, provided that the work is performed using alternative measures which provide effective protection.

This is not to suggest that worker protection must not be maintained during normal production operations. When employees carry out tasks during normal production operations with the machine or equipment energized, machine guarding must be provided in accordance with the provisions of OSHA's machinery and machine guarding standards (for instance, 29 CFR 1910.212 and 1910.219). The machine guarding standards require that workers be protected from hazardous moving parts of machinery during production. Dangerous moving parts in three basic areas require safeguarding: the point of operation; power transmission apparatus; and other moving parts. As a general rule, any machine part, function, or process which may cause injury must be safeguarded.

Paragraph (a)(2)(ii) of 1910.147 qualifies the exemption of normal production operations from the lockout/tagout standard, and provides further insight regarding covered activities:

> 1910.147(a)(2)(ii)
>
> Servicing and/or maintenance performed during normal production operations are subject to the standard only if:
>
> A) an employee is required to remove or bypass a guard or other safety device; or
>
> B) an employee is required to place any part of his/her body into an area on a machine or piece of equipment where work is actually performed on the material being processed (the "point of operation") or where an associated danger zone exists during a machine operating cycle.

Differentiating between service and/or maintenance and normal production operations is not always easy. For example, certain tasks such as minor, repetitive adjustments and simple tool changes are considered normal production operations when they are integral to the use of the equipment, and when they do not increase the risk of injury to employees. Other activities such as lubrication, cleaning, and unjamming, although integral to the use of the equipment, would be considered service or maintenance if employees are exposed to greater or different hazards than those encountered during normal production operations. Generally, if activities could expose employees to the unexpected energization or startup of the equipment or release of hazardous energy, and result in injury, they should be considered service and/or maintenance, and lockout/tagout must be implemented.

Essential Tasks

There are five essential tasks associated with launching a lockout/tagout program, including:

- performance of a workplace survey and hazard assessment to identify the MEP with hazardous energy, and the tasks that may expose workers to that hazardous energy;
- drafting a written program;
- purchase and provision of protective hardware and appliances;
- development of machine-specific energy control procedures; and
- employee training.

The completion of each of these tasks will serve as milestones in the implementation process. Each of these five tasks will be addressed in order. A sixth task, program review and revision, is associated with *Periodic Inspections*, and is addressed in Chapter 10, *Other Lockout/Tagout Requirements and Considerations.*

Workplace Survey and Hazard Assessment

The first step is to survey the workplace to identify the MEP that utilize hazardous energy, since the lockout/tagout standard applies to MEP that could accidentally energize, start up, or release stored energy and cause injury. It is also important to assess the types and magnitudes of the energy that must be controlled. Four general forms of energy are used alone or in combination to power machines and equipment and may be encountered during servicing or maintenance:

Mechanical Energy. The (kinetic) energy associated with moving parts of mechanical systems. Mechanical energy can be manifested in linear or rotational mechanical motion.

Potential Energy. Stored energy that can be released as hazardous mechanical energy. Potential energy is associated with pressure in pneumatic, hydraulic, and vacuum systems, with springs under tension or compression, and with gravity (for instance, raised loads).

Electrical Energy. From generated electrical power, sources of static electricity, or electrical storage devices (for instance, batteries, capacitors).

Thermal Energy. The energy resulting from high or low temperature. Thermal energy can result from mechanical work, radiation, chemical reaction, or electrical resistance.

For purposes of the survey, all forms of energy with magnitude sufficient to cause injury should be considered.

Table 2.2: Examples of covered machines, equipment, and processes

Agitators	Excavating machinery
Mixers	Earth moving equipment
Compactors	Fillers
Conveyors	Extruding, injecting, and molding machinery
Crushers and pulverizers	Packaging, bottling, and wrapping equipment
Forming rolls and calenders	Presses, shears, and forges
Pumps	Electrical apparatus
Compressors	Balers, harvesters, and reapers
Conveyors	Robots
Drillers and augers	Piping systems
Saws and cutters	Vehicles and equipment
Elevators and hoists	

The Written Energy Control Program

To put the lockout/tagout program into practice, supervisors and employees need specific guidance on what is expected of them. The written program establishes a given company's general policies and procedures for implementing lockout/tagout, sets specific performance requirements for employees, and provides the mechanism for regulatory compliance. The written energy control program should:

- address the program's purpose, scope, and application;
- define key terms;
- prescribe the responsibilities of managers, supervisors and employees for implementing program elements; and
- outline general lockout/tagout rules and procedures.

To achieve management buy-in, the draft written program should be circulated to affected managers and supervisors for review and comment; appropriate comments should be incorporated into the final document. The final version of the written program should be forwarded to members of senior management for concurrence and sign-off. The signature of senior managers will attest to their support for the program. The written program can then be adopted and issued to all employees in final form for implementation.

The written program should establish the general rules and procedures for implementing the lockout/tagout program in the workplace. Two of the most basic rules of lockout/tagout, for example, are: 1) all machines, equipment, and processes must be locked and/or tagged out whenever necessary to protect against accidental or inadvertent operation when such operation could cause injury to personnel; and 2) no one shall attempt to operate or energize any machine, equipment, or process when it is locked and/or tagged out. Other rules should address preparation for and application of lockout/tagout, verification of isolation, group lockout/tagout, transfer of lockout/tagout protection, energized testing and positioning, release from lockout/tagout, etc.

The overall written program should outline, in general terms, the personnel, activities, and energy sources covered by the program. For example, covered employees should include all permanent, temporary, and contract employees who:

- service or maintain equipment in which the unexpected energization or start-up could cause injury;
- operate equipment that is being serviced or maintained; or
- work in an area in which a lockout or tagout system is being implemented.

Activities covered by the written program include servicing, maintaining, repairing, adjusting, cleaning, constructing, erecting, installing, inspecting or operating machines, equipment, or processes where the employee faces a risk of injury. The policy should also state its applicability to all sources of hazardous energy, including potential, mechanical, electrical, thermal, hydraulic, pneumatic, and chemical energy, and radiation.

The responsibilities of managers, supervisors, and employees for implementing various aspects of the program should be clearly defined in the document. Management is generally responsible for:

- drafting, periodically reviewing, and updating the written program;
- identifying the employees, machines, equipment and processes included in the program;
- ensuring that employees are properly trained;
- providing the necessary protective equipment, hardware, and appliances; and
- monitoring/measuring conformance with program requirements.

Supervisors are generally responsible for:

- distributing protective equipment, hardware, and appliances and ensuring its proper use by employees;
- ensuring that equipment-specific procedures are established for the machines, equipment, and processes in their area;
- ensuring that only properly trained employees perform service or maintenance under lockout/tagout; and
- ensuring that employees under their supervision follow established lockout/tagout procedures.

Employees are generally responsible for:

- assisting in the development of equipment-specific procedures;
- following the procedures that have been developed; and
- reporting any problems associated with those procedures.

An example of a written lockout/tagout program is provided in Appendix D.

Protective Hardware and Appliances

To perform service or maintenance safely and effectively under lockout/tagout, employees need to be provided with protective equipment, hardware, and appliances for preventing unwanted energy transfer. Examples of the devices that may be used for isolating, securing, or blocking of machines or equipment from energy sources include locks, tags, chains, wedges, blocks, blinds, pins, fasteners, and other hardware. Lockout and tagout devices must meet specific requirements for design and construction, and in general must be durable and standardized. The majority of needed hardware and appliances are available from original equipment manufacturers and third-party vendors. In rare circumstances, devices may have to be custom fabricated.

Energy Control Procedures

In addition to the written policy's general rules and procedures, more detailed equipment-specific "energy control procedures" (ECPs) need to be prepared. Section (c)(4) of the lockout/tagout standard requires that ECPs be "developed, documented, and utilized for the control of potentially hazardous energy when employees are engaged in the activities covered." The ECPs supplement and work in conjunction with the general provisions of the overall written program. In general, ECPs should describe the:

- machine, equipment, or process to be locked/tagged out;
- intended use of the procedure or type of service/maintenance to be performed;
- types and magnitude of hazardous energy associated with the machines, equipment, or processes (MEP);
- type and location of the isolating device(s);
- involved personnel;
- specific procedures and sequences for de-energizing, verifying, and locking/tagging out the MEP; and
- specific steps for re-energizing the machine and restoring it back to normal operation.

Key to developing good ECPs is a thorough understanding of the service and maintenance tasks that are performed on MEP. In addition, if there are changes in the equipment, method or location of isolation, the energy sources, or involved employees, ECPs need to be modified as necessary to reflect those changes. Energy control procedures are addressed in further detail in Chapter 8.

Employee Training

The lockout/tagout standard requires that employees receive training prior to assignment to ensure that they understand the purpose and function of the energy control program and have the knowledge and skills required for the safe application, use, and removal of energy controls. The level of training provided must be commensurate with employee's respective job assignments. For example, different levels of training are prescribed in the standard for "affected" vs. "authorized" employees. An "affected employee" is an employee whose job requires him/her to operate or use a machine or equipment on which servicing or maintenance is being performed under lockout or tagout, or whose job requires him/her to work in an area in which such maintenance is being performed. Affected employees must be instructed in the purpose and use of the lockout/tagout policy and procedure, and the prohibition relating to attempts to restart or re-energize MEP which are locked out or tagged out. An "authorized employee" is a person who locks out or tags out machines or equipment to perform the servicing or maintenance on that machine. Authorized employees must receive training in the recognition of applicable hazardous energy sources, the types and magnitude of the energy available in the workplace, the methods and means necessary for energy isolation and control, and the means of verification of energy control.

Retraining is required whenever there is: a change in job assignment; a change in machines, equipment, or processes that present a new hazard; or a change in the energy control procedures or revision of control methods. Additional retraining is also required whenever a periodic audit or supervisory observation of the energy control procedures indicates deviation from or inadequacies in an employee's knowledge or use of the procedures.

Employee training is discussed in more detail in Chapter 9.

Key Success Factors

While meeting the basic provisions of the lockout/tagout standard may suffice for compliance purposes, additional elements are needed for the program to be successful, including:

Visible Management Leadership. Management must clearly demonstrate the importance of the lockout/tagout program by involvement, support, and example. Adequate human and financial resources must be devoted to the program. Responsibilities for implementing the program should be well defined, and personnel must be empowered to carry out those responsibilities. Managers should follow up to ensure that all requirements are properly executed. Protective appliances (for instance, locks, tags, blocks, pins, wedges, and other lockout devices) must be provided and readily available.

Employee Participation. Employees should participate in the development of the energy control procedures that affect them and be trained in their use. Such active employee involvement enhances understanding of the procedures, and fosters a sense of "ownership." Communication with and feedback from employees should be sought continuously to identify deficiencies in the lockout program, as well as opportunities for improvement.

Enforcement and Accountability. Injuries associated with improper lockout are often serious or fatal. Personnel must therefore be held accountable and understand the consequences for violating lockout/tagout procedures. Discipline should be initiated when lockout/tagout violations occur to bolster enforcement. Many companies have disciplinary policies in which progressive discipline steps are skipped, and actions "up to and including discharge" are imposed when safety violations such as failing to follow lockout/tagout occur. The risks associated with non-adherence to lockout procedures are simply too high to tolerate.

Chapter 3

Methods of Energy Control

Successful application of energy control requires an understanding and awareness of hazardous energy types and magnitudes, and the methods, techniques, and devices used to control or isolate that energy. Energy control can range from simple procedures involving only a few steps for single-energy source machines, to complex isolation involving multiple energy sources, multiple machine components, and numerous personnel. Several factors must be considered in deciding what method of energy control should be implemented for a particular application, including: the energy level; the access or proximity to the energy source necessary to accomplish the work; and the effectiveness of the methods chosen in providing employee protection. The methods of energy control used must match the hazards and complexity of the equipment involved, and the type of service or maintenance to be performed. Several different means of controlling energy may be needed simultaneously to control a particular operation.

The first step in controlling hazardous energy so that employees can perform service or maintenance safely is to shut down the machine, equipment, or process. Simply shutting off a machine, equipment, or process, however, may not completely control the hazardous energy. Even after shutdown, residual energy may still be present in sufficient quantity to cause unanticipated movement (and subsequent injury) if released, and may need to be controlled. For example, residual energy may be present in the form of moving components, gas, or liquid systems under pressure, springs under tension or compression, the force of items that have become jammed, or the energy that is stored in machine, equipment, or system components due to their position (for example, potential energy due to gravity). In addition, even if the machine, equipment, or process has been shut down and residual energy dissipated, an accident could still occur as a result of unexpected start up (activation) due to human error. For example, another employee could accidentally restart the machine, equipment, or process at the controls without realizing that the servicing employee is exposed. An accident could also occur if the servicing employee himself causes a short in a control circuit or accidentally turns on a system controller such as an electrical control switch or valve and inadvertently activates the machine.

For the majority of machines and equipment, therefore, several additional steps are necessary to effectively control hazardous energy:

- All energy sources identified as hazardous must be isolated or blocked at their points of control (energy isolating devices);

- All energy isolating devices that are needed to control the energy to the machine must be secured (locked and/or tagged out) to hold them in a "safe" or "off" position for the duration of the service or maintenance task; and
- All potentially hazardous stored or residual energy must be relieved, dissipated, disconnected, restrained, or otherwise rendered safe.

In achieving effective energy control and compliance with the lockout/tagout standard, it is important to ensure that all energy isolation points have been identified and isolated, and that the appropriate lockout/tagout device has been applied to provide protection against the potential release of hazardous energy.

Lockout/Tagout Terms

In the following pages, specific energy types and their control are discussed in greater detail. First, however, it is essential to be familiar with some important definitions that clarify the meaning, intent, and purpose of certain terms used in the lockout/tagout standard. Other important terms are defined in the glossary provided at the end of the book.

Capable of being locked out. An energy isolating device is capable of being locked out if it has a hasp or other means of attachment to which, or through which, a lock can be affixed, or it has a locking mechanism built into it. Other energy isolating devices are capable of being locked out if lockout can be achieved without the need to dismantle, rebuild, or replace the energy isolating device or permanently alter its energy control capability. A chain and lock combination or a lockable valve cover can be used to meet the definition of "capable of being locked out."

Energy source. Any source of electrical, mechanical, hydraulic, pneumatic, chemical, thermal, or other energy.

Energy isolating device. A mechanical device that physically prevents the transmission or release of energy, including but not limited to the following: A manually operated electrical cir-

Figure 3.1. Energy isolating devices do not include push buttons, selector switches, or other control circuit type devices.

cuit breaker; a disconnect switch; a manually operated switch by which the conductors of a circuit can be disconnected from all ungrounded supply conductors and, in addition, no pole can be operated independently; a slide gate; a line valve; a block; and any similar device used to block or isolate energy. The term does *not* include a push button, selector switch, or other control circuit type devices (see Figure 3.1). Energy isolating devices are the primary tool for pro viding protection under the standard; they are the mechanism that prevents the transmission or release of energy and to which locks or tags are attached. "Isolating" a machine or piece of equipment from an energy source means using an energy isolating device to interrupt the flow of energy from the means of transmission of the energy to the machine. Examples of energy isolating devices are provided in Figure 3.2.

Energized. Energized means connected to an energy source, or containing residual or stored energy.

Lockout. The placement of a lockout device on an energy isolating device in accordance with an established procedure, ensuring that the isolating device and the equipment being controlled cannot be operated until the lockout device is removed.

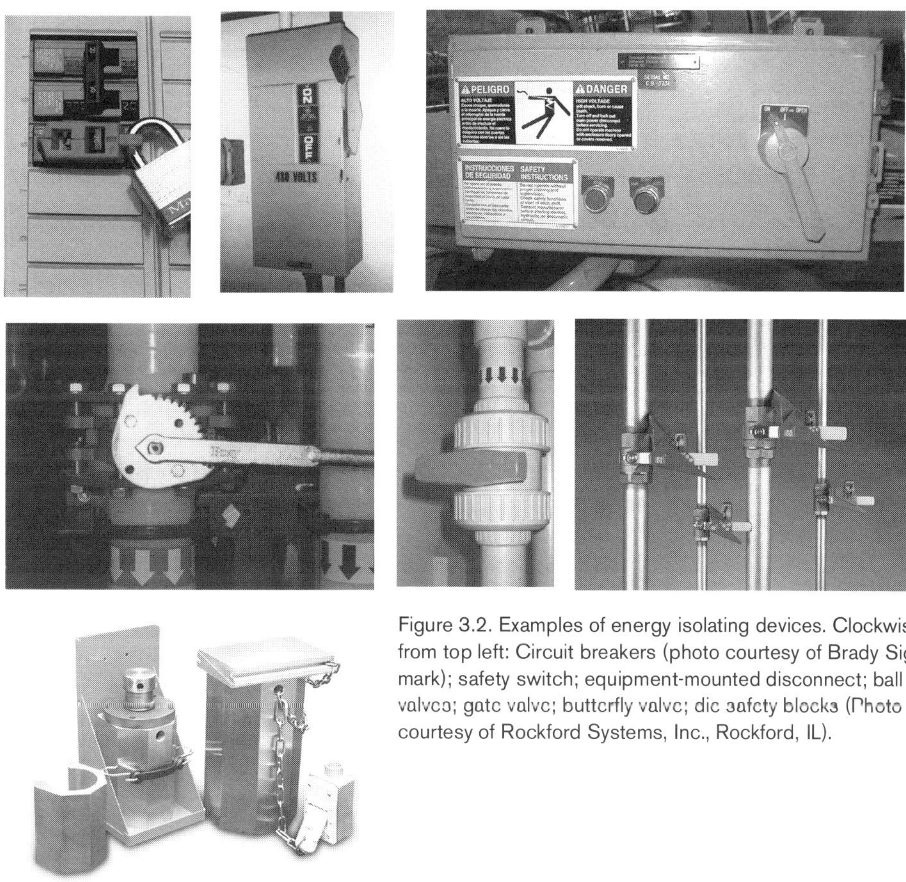

Figure 3.2. Examples of energy isolating devices. Clockwise from top left: Circuit breakers (photo courtesy of Brady Signmark); safety switch; equipment-mounted disconnect; ball valves; gate valve; butterfly valve; die safety blocks (Photo courtesy of Rockford Systems, Inc., Rockford, IL).

Lockout device. A device that uses a positive means such as a lock, either key or combination type, to hold an energy isolating device in the safe position and prevent the energizing of a machine.

Tagout. The placement of a tagout device on an energy isolating device, in accordance with established procedure, to indicate that the energy isolating device and equipment being controlled may not be operated until the tagout device is removed.

Tagout device. A prominent warning device, such as a tag and a means of attachment, that can be securely fastened to an energy isolating device in accordance with an established procedure to indicate that the energy isolating device and the equipment being controlled may not be operated until the tagout device is removed (see Figure 3.3).

Energy Sources

Four main forms of energy may be present, alone or in combination, in machines, equipment, and processes: mechanical energy, potential energy, electrical energy, and thermal energy. In general, an energy source is "hazardous" and must be controlled if it is capable of causing injury. Where multiple energy sources are encountered, each must be effectively controlled.

Electrical Energy

Electrical energy is the most common energy type used to power machines and equipment. This form of energy generally refers to generated electrical power and sources of static electricity, but may also be encountered with electrical storage devices such as batteries and capacitors. Electricity is used as a primary energy source to power motors, pumps, drives, controls, and other devices in a great deal of industrial machines, equipment, and processes. Workers in many different industries are exposed to electrical hazards in the performance of their duties. Contact with electrical energy can result in serious injury, including: electrocution (fatal), electric shock, burns, and falls that are caused as a result of electric shock. Serious injury can also occur if uncontrolled electrical energy causes the inadvertent activation or start up of machines or equipment while a worker is exposed. Numerous injuries due to electric shock and more than 100 electrical fatalities occur in general industry each year. To prevent injury, electrical energy sources must be de-energized and locked/tagged out whenever covered service or maintenance tasks are performed.

Figure 3.3. Tagout device. (Courtesy of Brady Signmark.)

Electricity is created by the flow ("current") of electrons through a conductor. Voltage, measured in volts, is the force that causes the electricity to flow, and is typically used to describe the magnitude of the electrical energy.

Current is a measurement of the amount of electrons flowing through a conductor and is expressed in amperes (amps). Resistance is anything that impedes the flow of electricity through a conductor and is measured in ohms. Voltage, current and resistance can be related mathematically using Ohm's Law (voltage = current × resistance). Two types of electrical current are used—alternating current (AC) and direct current (DC). In AC systems, the current flows back and forth ("cycles") through the conductor, first in one direction, then the other. In DC systems, current flows only in one direction. AC is most widely used because it is possible to step the current/voltage up or down to the level needed by the equipment by using a transformer. Common energy requirements include 120/240 volts nominal for simple motors and household appliances, 277 volts for lighting systems, and 480 volts for many industrial machines. High voltage systems (greater than 600 volts nominal, such as those encountered in power generation, transmission, and distribution) commonly utilize 600, 13,800, or 118,000 volts.

Large industrial facilities typically receive power from a utility or power company at an electrical substation. The substation contains transformers, switchgear, and circuit breakers that control and distribute (generated) power throughout the facility via feeder circuits. Feeders provide power to distribution stations (panelboards, switchboards, and motor control centers), from which (utilization) power is distributed through branch circuits to machines, equipment, light fixtures, receptacles, etc. Motor control centers often include a transformer to further step down power, and an oil-filled or air-break disconnect switch that can be used to interrupt power.

The motor control centers and panelboards are provided with individual circuit breakers (or fuses) that provide overcurrent protection for the branch circuits and that can also be used to interrupt power. This system of power distribution is similar to the way that limbs and branches fork off the main trunk of a tree; power distribution circuits for various parts of the system branch off the main supply of electricity; each "fork" in the wiring is represented by a switchboard or panelboard containing the circuit breakers that supply feeders (limbs) and branch circuits (branches).

In addition, a local disconnect or safety switch is usually required to be installed in sight from industrial machines, motors, and appliances to provide a means of disconnecting the equipment from the electrical source. Smaller facilities may receive electricity from the utility or power company through a transformer which, in turn, feeds power to a motor control center or panelboard. Facilities may also have back-up power for systems and equipment, provided by generators and uninterrupted power supplies, and a transfer switch may be provided for automatically and/or manually switching to back-up power in the event of a power outage.

The lockout/tagout standard (29 CFR 1910.147) addresses practices and procedures that are necessary to de-energize machinery or equipment and to prevent the release of potentially hazardous energy while maintenance and servicing activities are being performed. Although the rule is related to electrical energy, it specifically excludes "exposure to electrical hazards from work on, near, or with conductors or equipment in electric utilization installations, which is covered by Subpart S of 29 CFR 1910."

The electrical safety standards of Subpart S rather than the lockout/tagout standard, therefore, cover the hazards of electrical utilization. While the standards of Subpart S mainly address electrical design, construction, and installation, they also include a standard for Electrical Safety-related Work Practices (1910.333). This standard prescribes requirements for work on electric equipment, including provisions for de-energizing and locking/tagging out electrical circuits. As an alternative to complying with the locking/tagging requirements set forth in 1910.333, compliance with the lockout/tagout (1910.147) standard's provisions are considered acceptable by OSHA. While the lockout and tagging provisions of these two standards may differ in wording, the basic concepts of de-energizing and locking/tagging out machines and equipment prior to working on them remain the same.

Although the specific methods used to lockout/tagout machines and equipment that use electrical energy vary depending on the installation, for industrial equipment, the first step is generally to shut the machine down by the normal stopping means. Generally this is done at operating controls such as push buttons or selector switches (for example, "on/off" or "start/stop") which themselves use electrical circuitry. Next, the machine must be isolated (disconnected) from all external electrical power sources at the appropriate electrical isolating devices. Workers should be aware that there can be several sources of electrical power to machines, and that lockout/tagout at several points of control can be required. For example, electrical energy can be used both for powering (e.g. motors and pumps) and controlling (e.g. motor controllers and electrically-operated automatic valving). Separate electrical sources may be present for controls and power respectively, and one or both of these sources may need to be isolated and locked/ tagged out, depending on the service task to be performed.

Figure 3.4. Lockout/tagout at circuit breaker. (Photo courtesy of Brady Signmark)

Figure 3.5. Lockout/tagout at disconnect switch.

Circuit Breakers and Disconnects

The primary devices for controlling and isolating electrical energy are circuit breakers and disconnect switches. These provide the mechanism that prevents the transmission or release of electrical energy and are the devices to which locks and/or tags are applied. For the majority of industrial machines and equipment, electrical energy is isolated by opening (placing in the off position) and locking/tagging out, either the local disconnect or the main disconnect (back at the panel or motor control center). The vast majority of this equipment is capable of being locked out in this fashion (see Figures 3.4 and 3.5).

Control circuit devices such as push buttons, selector switches, emergency stops, and so on play an important part in shutting down machines and equipment prior to service or maintenance. Such mechanisms, however, are generally not adequate for, and may not be used as, the sole means for de-energizing circuits or equipment. The intent of the lockout/tagout standard was not to accept such mechanisms as energy isolating devices, and in fact, OSHA's definition of energy isolating device specifically excludes push buttons, selector switches, and other control circuit type devices.

Fuses

Fuses are also devices that are used for energy isolating and overcurrent protection. Their removal alone, however, is not adequate for preventing the transmission or release of electrical energy (they could be inadvertently reinstalled), and must be supplemented with additional or alternative measures to ensure that electrical energy is not restored while workers are exposed. This may be accomplished, for example, by locking/tagging out the fuse box itself after the fuse is removed. If the fuse box cannot be padlocked and tagging procedures alone will be used, methods which ensure full employee protection, (protection equivalent to that provided by locking/tagging out a breaker or disconnect) must be used. In addition to removing the fuse,

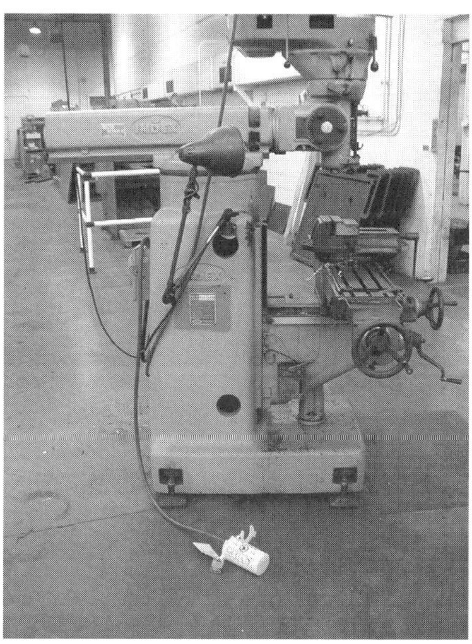

Figure 3.6. Plug lockout device used on cord and plug connected milling machine.

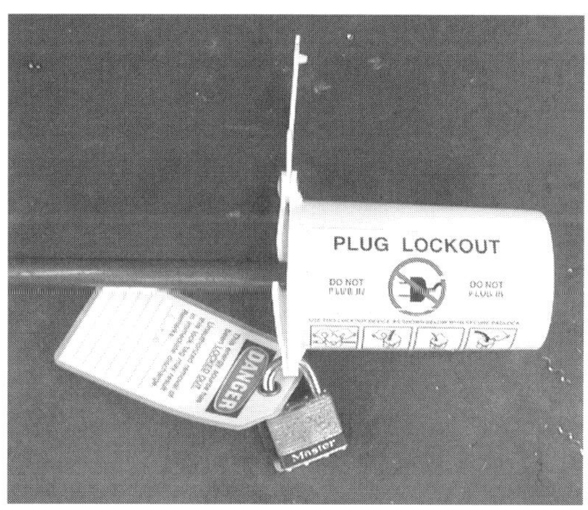

therefore, additional steps may be needed, including blocking of a controlling switch, opening of an extra disconnecting device, stationing a person at the fuse block to ensure that fuses are not reinstalled, and so on (see "Alternative Measures" on page 42). In addition, tags should be affixed to removed fuses (and in other locations as necessary) to warn against reinstallation until the tag is removed by the servicing employee who affixed it.

Cord and Plug Connected Equipment

Many machines, equipment, and appliances receive power through flexible electrical cords and plugs. Examples include grinders, polishers, mixers, heaters, saws, drill presses, and other equipment. Cord and plug connected equipment can expose service or maintenance personnel not only to hazardous electrical energy, but to other energy types as well. These hazards must be controlled by unplugging the equipment from the energy source, along with any other necessary measures so that service or maintenance can be performed safely (see Figure 3.6).

With respect to the locking/tagging out of cord and plug equipment, an exemption from coverage under the lockout/tagout standard is provided for this equipment in 1910.147(a)(2)(iii)(A).

> 1910.147 (a)(2)(iii)
>
> This standard does not apply to the following:
>
> (a)(2)(iii)(A) Work on cord and plug connected electric equipment for which exposure to the hazards of unexpected energization or start up of the equipment is controlled by the unplugging of the equipment from the energy source and by the plug being under the exclusive control of the employee performing the servicing or maintenance.

This exemption only applies, however, when these conditions are met. "Under the exclusive control of the employee" means that the authorized employee is able to prevent the equipment from becoming re-energized while it's being serviced. The plug is under the exclusive control of the employee if:

- it is physically in the employee's possession; or
- it is within arm's reach and in the employee's line of sight; or
- the employee has affixed a lockout/tagout device to it in accordance with the standard.

If the cord extends beyond reach of the servicing employee and the plug is out of sight, there is potential for another employee to accidentally restart the machine by plugging it in without realizing that the servicing employee is exposed. In these situations, the employee must affix a lockout/tagout device to the plug. The most prudent approach is to place a lock and tag together on the plug lockout to physically prevent it from being plugged in. In addition, the company's energy control program should specify the acceptable procedure for handling cord and plug connected equipment.

A common citation issued by OSHA for violations of the electrical requirements of Subpart S of the General Industry Standard involves the use of flexible cords or cables as a substitute for the fixed wiring of a structure. Pursuant to OSHA (29 CFR 1910.305(g) and NE (Article 400-8) requirements, flexible electrical cords may only be used for:

- Pendants;
- Wiring of fixturcs;
- Connection of portable lamps or appliances;
- Elevator cables;
- Wiring of cranes and hoists;
- Connection of stationary equipment to facilitate their frequent interchange;
- Prevention of the transmission of noise or vibration;
- Appliances where the fastening means and mechanical connections are designed to permit removal for maintenance and repair; or
- Data processing cables approved as a part of the data processing system.

Inspections often reveal instances of improper use of flexible cord to power industrial machinery. Before using the lockout/tagout standard's cord and plug exception, therefore, it's a good idea to first review the use of the flexible cord for the installation, and ensure that it is approved and suitable for the conditions of use and location. If necessary, the machine should be rewired using an approved wiring method, and an approved disconnecting means provided. The lockout/tagout standard's provisions would then be applied.

Batteries and Capacitors

Stored electrical energy is encountered in devices such as batteries and capacitors. Even with the main power off, circuits that use these devices can retain sufficient energy to reenergize electric circuit parts or cause inadvertent machine activation. This energy must be blocked or relieved when necessary to ensure that circuit parts cannot be accidentally energized.

A capacitor is an electrical device that stores an electric charge. It consists of two conducting plates of metal separated by an insulating material called a dielectric. Capacitance (C) is the ability to store an electric charge expressed in Farads (F). Other systems may employ a capacitor bank, which is a group of electrically connected capacitors. Some capacitors can store extremely high voltages. If the stored electric energy of capacitors or other high-capacitance elements might endanger personnel, they must be discharged (for example, short-circuited and grounded). For new installations, capacitors must have an automatic means for discharging themselves (see Article 460-6 of the National Electrical Code).

Batteries are devices that use a chemical process to create electricity. In direct current (DC) systems, batteries may be used to provide power directly to utilization equipment. Other common uses of batteries include back-up/emergency power for critical equipment systems, uninterrupted power supplies, and start-up power for vehicles and equipment. Some battery systems may be equipped with breakers or disconnects that can be used to isolate and lock/tag out the electrical energy. In other systems, however, removal, tagging and/or locking of battery leads may be necessary. Often, a plug lockout device allows the battery lead to be locked out.

ı Energy

‚or kinetic) energy is manifested through motion, and is the energy associated with ‚rts of mechanical systems. Mechanical energy can be manifested in linear and rotaechanical motion, and through translation or oscillation. It is often encountered in transmission apparatus such as shafts, flywheels, sheaves, pulleys, belts, and clutches, ıt the point-of-operation of machines. Motion can also produce work, which in turn can ›duce changes in temperature, due to friction. This type of energy can be turned off, or left ›n. Figure 3.7 provides examples of various hazardous motion types.

Contact with dangerous moving parts can result in serious crushing, cutting, lacerating, amputation, or fracture injury to an employee's hand, arm, or other body part, or even death. Consequently, machines must be safeguarded in accordance with the OSHA machine guarding requirements of 29 CFR Subpart O to prevent workers from placing any part of their bodies into a hazardous machine area during normal production operations. During service or maintenance activities, however, safeguarding may be removed, bypassed, or disabled, or workers may otherwise be exposed to unexpected energization or start up, and machines and equipment must therefore be locked/tagged out to protect employees.

To control mechanical energy, the machine must first be shut down and isolated from its primary energy sources if necessary. The isolating devices for the primary energy sources must be locked/tagged out to prevent the machine from being reenergized. Any moving parts, such as flywheels, blades, gears, and shafts, or cam, concentric, and pivotal machine components, that coast or remain in motion after shutdown should be allowed to come to a complete stop. An accident that occurred in California illustrates the need to observe this precaution. A table saw was turned off, but the saw blade was still coasting and had not come to a complete stop. An employee began cleaning the machine, and his finger was amputated by the blade. Certain types of industrial machinery such as large metal-forming presses may have moving components (for example, flywheels) that take 10 or more minutes to stop, and in some cases brakes may be needed on the components to stop them.

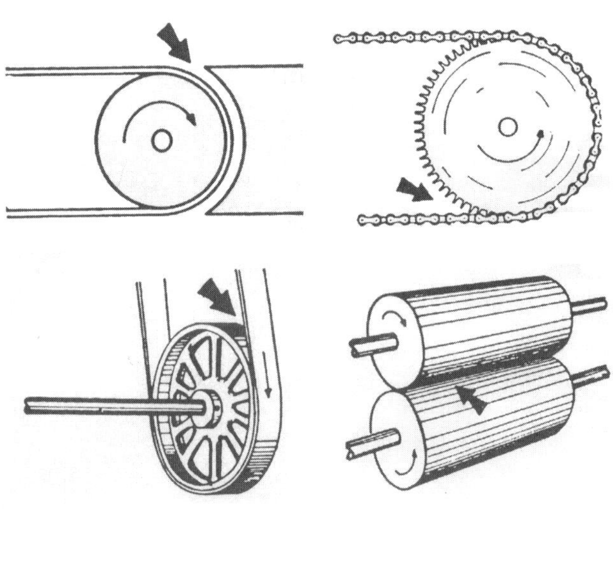

Figure 3.7. Examples of hazardous motion types. (Source: *Concepts and Techniques of Machine Safeguarding.* OSHA Publication 3067, 1992.)

In some cases, potentially dangerous machine member movement may be possible even after shutting the equipment down and locking out the primary energy sources. If potentially dangerous movement is possible, blocking with pins, blocks, bars, etc. or disengaging moving parts may also be needed in addition to shutting off the primary energy sources. The following case study illustrates this point. A maintenance employee was changing v-belts on an exhaust fan. He de-energized the fan before starting work, but did not block the blades of the fan to secure them from movement. Air movement in the duct caused the fan blades to turn, and his hand was caught and crushed in the v-belt drive.

Figure 3.8. Use of pins on palletizer to hold elevator in raised position.

Blocking or disengaging moving parts may also be needed if employees themselves could cause dangerous machine member movement. For example, if it is necessary for a worker to climb or move about mechanical linkages, the worker's weight may be sufficient to cause dangerous movements of the linkages.

Mechanical energy can also be stored as potential energy (for example, in suspended or jammed parts or springs), which, if released, could cause unexpected energization or activation of the machine. This stored energy must also be dissipated or restrained when necessary to prevent movement (see discussion on potential energy below).

Potential Energy

Potential energy is stored energy that can be hazardous if released. This type of energy is often not as obvious as incoming energy sources such as electric, hydraulic, or pneumatic. Potential energy in many forms can be stored in sufficient quantity to present a hazard. Examples include:

- Gravity, as in raised or suspended parts;
- Pressure (above or below atmospheric), as in hydraulic, pneumatic, or vacuum systems;
- Compressed or extended springs;
- Static electricity;
- Stored electrical energy, as in batteries and capacitors;
- Thermal energy due to residual heat or low temperature;
- Residual chemicals in piping systems that may cause thermal or pressure buildup; and
- Stored energy associated with buoyant objects that are anchored underwater.

The most effective way to prevent employee injury due to inadvertent movement of machine components is to dissipate any potential or residual energy to a non-hazardous level, or to use a restraining device such as a block or pin to prevent movement (see Figure 3.8). This can be done, for example, by: blocking materials or moving components to a stable, resting position; by moving components to a point at which springs are at a neutral (relaxed) state; and by bleeding down systems such as air, hydraulic, or steam lines that contain liquids or gases.

Potential Energy Due to Gravity

Potential energy due to gravity is a function of the height of an object such as a machine component above a reference plane. The reference plane is the place where the object would come to rest if it were to fall, slide, or roll—for example, to ground level. If the potential energy of a raised or suspended object becomes unrestrained, it can be converted into potentially injurious kinetic energy in the form of downward motion. Blocks, stands (such as those commonly used under raised vehicles), special brackets, pins, and wedges must therefore be used to hold the weight of raised or suspended materials and components and prevent them from falling, sliding, or rolling (see Figures 3.9 and 3.10).

Springs

Potential energy (stored mechanical energy) may also be found in springs under tension or compression. The magnitude of this energy depends on the size and construction of the spring, and varies with the load imposed on the spring, and its position. Springs may be encountered on presses, shears and cutters, vehicles, roll-up doors, dock levelers, and other equipment. Types of springs include helical coil springs (Figure 3.11), flat coil springs, torsion springs, and leaf springs.

The energy associated with spring-loaded systems can be significant, and serious injury could result if this energy is not anticipated and controlled. The method used to control this energy may vary depending on the type of equipment and the task involved. For example, moving machine components to a position in which the spring is in a relaxed state may be sufficient to dissipate the potential energy. Special tools and devices such as clamps, caps, and retainers may be needed to compress, relax, or secure the spring, and special proce-

Figure 3.9. Stands used under floor hoist.

Figure 3.10. Die blocks on power press. (Source: *Concepts and Techniques of Machine Safeguarding*, OSHA Publication 3067, 1992.)

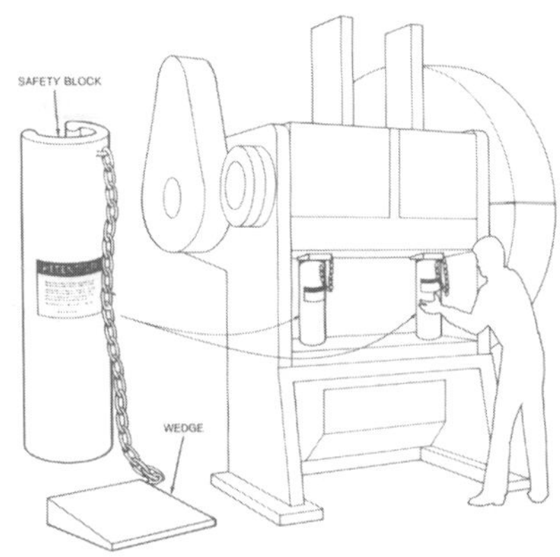

Figure 3.11. Helical coil spring.

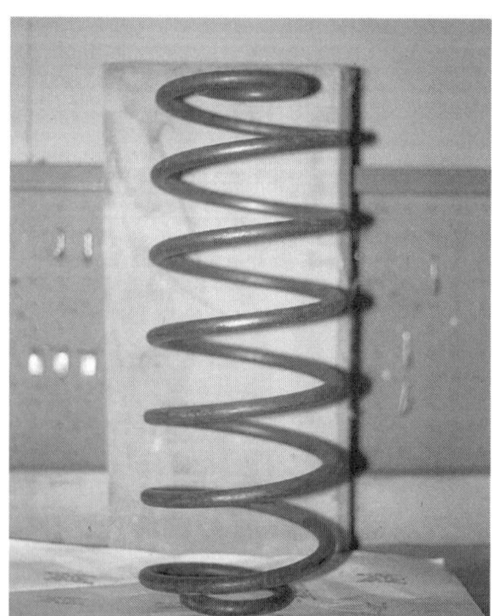

dures may be needed to ensure the safe control of spring energy.

Energy Control in Piping Systems

Piping systems, including pipes, lines, tubes, hoses, and ducts, are used to transport a wide variety of materials, including gases, liquids, semi-liquids, and particulates. For example, process and utility piping distributes chemicals, steam, water, compressed air, and other products throughout industrial facilities. Auxiliary devices used in conjunction with piping systems include pumps, tanks, pressure vessels, reservoirs, accumulators, valves, gauges, and regulators. Residual energy may be present in these systems in the form of stored or trapped pressure, vacuum, or residual chemicals that could cause thermal or pressure buildup. When necessary, this residual energy must be relieved and the system isolated so that service or maintenance can be performed safely.

Various types of pumps are used to move materials through piping systems. Pumps may be driven by electric motors, steam turbines, or internal combustion engines. Their size and construction depends on the type of service, pressure, temperature, and nature of the materials.

Shutting off and locking/tagging out the main power source to the pump is a customary first step in isolating the system in preparation for maintenance. A pump will always have an inlet (designated "suction") and an outlet (designated "discharge"); isolation at one or both sides may need to be accomplished depending on the nature of the service task to be performed (see Figure 3.12).

Figure 3.12. Basic pump configuration. (Source: *A Guide to Safety in Confined Spaces*, NIOSH Publication No. 87-113).

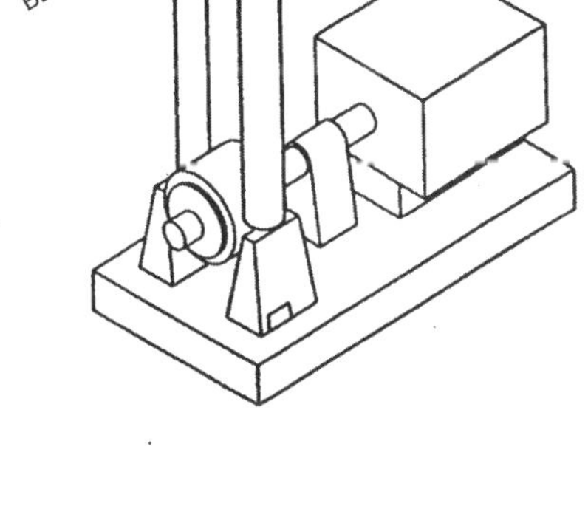

Various units are used to describe the magnitude of energy in pressurized systems. In the English system, fluid pressure and vacuum is generally measured in pounds per square inch (psi) and pounds per square foot. In the SI system, fluid pressure is measured in pascals (Pa) or kilopascals (kPa). Other units of pressure include bars, millibars, atmospheres, inches or feet of water, millimeters or inches of mercury, and torr.

Valves are the main devices used for the isolation and relief of energy in piping systems. Valves of various types are used for sectioning, directional and flow control, draining, and pressure control and relief. Valves can be manually or automatically operated, and the type of valve used depends on their operating purpose. Commonly used valves include gate valves, ball valves, globe valves, and butterfly valves. Generally, these valve types are capable of being locked out, since they incorporate an integral means for the attachment of a lock or hasp, or can otherwise be fitted with a chain or lockout device, to which a lock or hasp can then be attached (see Figures 3.13 and 3.14).

Other valve types include plug, rotary, check, bypass, and relief valves. The shape or location of some valves may make them difficult to lock out. With systems that use automatic valving that is difficult or impossible to lock out, installation of additional line valves that are capable of being locked/tagged out should be considered to enhance isolation capability. These additional line valves could be left open during normal production and closed only when needed for service or maintenance. As a precaution, control power for automatic valving may also need to be shut off (for example, to prevent damage due to lack of prime).

Figure 3.13. Lockout/tagout of gate valve. (Photo courtesy of Brady Signmark)

Figure 3.14. Although some valves are not designed with an integral means of being locked, they can be secured with chains, cables, blocking braces, or wedges, which can then be locked. (Photo courtesy of Brady Signmark)

Effective isolation of piping systems may be accomplished by closing and locking/tagging the appropriate control and section valves, or by other means such as blanking or blinding, or misaligning or removing sections of lines, pipes, or ducts. Pressure relief is typically accomplished by opening the appropriate drain or relief valves to bleed the system of pressure, or by otherwise venting fluids, air, or other gases from the system. Many systems incorporate special bleeder valves on process vessels or lines for this purpose (see Figure 3.15).

Often, a "double block and bleed" system is used to isolate and remove residual energy from a piping system. A double block and bleed is the closure of a line, duct, or pipe by closing and locking/tagging two in-line valves (upstream from the work section) and by opening and locking or tagging a drain or vent valve in the line between the two closed valves. When necessary, the drain or vent valve should remain open for the duration of the service or maintenance task to prevent the reaccumulation of pressure or other energy. In this case, it may also be appropriate to lock and tag the drain valve in the *open* position. Double block and bleed valves must be regularly checked to ensure that no through leakage can occur.

In addition to valves, blanks or blinds—when properly bolted in place—are also acceptable devices for achieving energy isolation in piping systems. Blanking or blinding is accomplished by

Figure 3.15. Bleed valves.

Figure 3.16. Method of blanking piping system. (Source: *A Guide to Safety in Confined Spaces*, NIOSH Publication No. 87-113.)

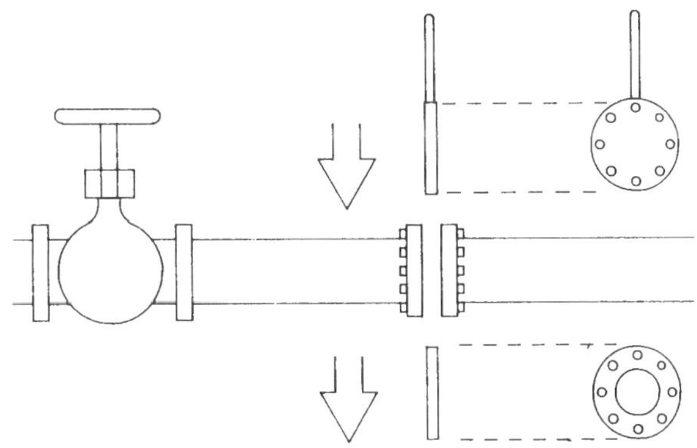

inserting a solid plate (such as a "spectacle" or "skillet" blind) between the flanges of a pipe, line, or duct joint, and bolting it in place (see Figures 3.16 and 3.17). The blind completely covers the bore and is capable of withstanding the maximum pressure of the pipe, line, or duct without leakage. Placement of the blind ensures that no liquids or gases will pass through that point if the system is accidentally activated. In high-risk situations, where valves alone cannot be relied on as a means of isolation because they can unexpectedly leak gases, vapors, or fluids, blanking may be used as an additional measure to supplement valve lockout. In other situations—for example, where no valve is provided for energy isolation—line blanking may be the only way to effectively isolate the piping system. Before installing blinds or blanks, care must be used to ensure that residual pressure in the system is bled off. In addition, since the lockout/tagout standard requires the lockout devices to identify the person that affixes them, a supplementary tag should be affixed to the bolted blinds and flanges.

Figure 3.17. Blank installed on vessel flange.

In some cases, relief valves may not be present in the piping section to be serviced, and special line breaking or pipe opening procedures may be needed for depressurizing or draining the system without endangering the employees. Line breaking is the intentional opening of a line, pipe, or duct that is or has been carrying flammable, corrosive or toxic material, an inert gas, or any fluid at a volume, pressure, or temperature capable of causing injury.

An example of a pipe opening procedure is as follows:

Pipe Opening Procedure

Pipe opening includes the breaking, unbolting, or unscrewing of flanges and fittings, removing valves or sensors, or cutting of the pipe or equipment.

> **Note** A hot work permit may also be required if cutting, grinding, or other activities involving open flames or producing heat or sparks are conducted.

1. Identify the contents of the piping system. Consult piping and instrument diagrams (PIDs), pipe markings and labels; trace the piping back to the source if necessary.
2. Review Material Safety Data Sheets (MSDs) for hazards and safe handling procedures for any substances that may be encountered.
3. Inform all affected persons in the area of the pipe opening. Clear affected area of unauthorized personnel, and remove or cover materials and equipment that could be affected by the possible release of piping system contents.
4. Ensure that a safety shower/eyewash or other facility for quick drenching of the eyes and body is nearby, functional, and easily accessible.
5. Cordon off a 10-foot radius around the pipe or equipment with rope or barrier tape and restrict access to the area.
6. Station an attendant at a safe distance to observe the pipe opening and assist in getting any splashed employee to the safety shower/eyewash.
7. If necessary, protect nearby drains, catch basins, and soil by covering with plastic sheeting resistant to the chemicals in the pipe or equipment, or by diking or damming, and place a container under the pipe or equipment to catch any released liquid.
8. Close and lock/tag out all line valves leading into or from the pipe or equipment. Lock/tag out all pumps, compressors, agitators, etc. connected to the pipe or equipment.
9. Put on appropriate personal protective equipment (PPE) such as splash-proof goggles, face shield, chemical-resistant boots, coveralls, and gloves.
10. Depressurize, drain or purge the system using existing drains or vents, if possible. Allow sufficient time for venting or draining. Confirm the completion of draining or venting if possible by checking available pressure or level gauges.
11. If unable to drain or vent, or if unable to confirm draining or venting, use shielding such as plastic tarpaulins or rubber matting to provide a barrier between the employee and the pipe or equipment to divert splashes.
12. Open the side of the pipe or equipment away from the employee first so that any unexpected splash or release flows away from him or her (do not stand in the line of fire). Slowly loosen bolts on fittings or flanges, and be alert for escaping pressure or materials.
13. In systems containing liquids, next loosen and remove the bottom bolts. Catch any dripping or leakage that occurs. Remove the remaining bolts.

14. Separate flanges or fittings carefully. When removing pipes or equipment, tilt if necessary to completely drain. Use care to ensure that residual contents are not accidentally released from openings. Drain residual contents into an appropriate container, as necessary.

15. Flush, neutralize, or decontaminate pipe or equipment if necessary. Collect and dispose of all waste materials properly.

16. Before returning the piping system to service, verify that the equipment is in a condition to be safely energized. Verify that all non-essential items, such as tools, have been removed, that the equipment is operationally intact, and that controls are in a safe "off" position. Before recharging/reenergizing the system, visually verify that all persons are safely positioned.

Figure 3.18. Well-marked piping systems.

Identification of Piping Systems

Although marking or labeling of piping systems and valves is not required by the lockout/tagout standard, it is nonetheless an important aid to workers in properly identifying piping system contents. Piping systems should also be identified and marked in accordance with ASME A13.1-1996—Scheme for the Identification of Piping Systems. Apply the legends close to valves or flanges and adjacent to changes in direction, branches, and where pipes pass through walls or floors, and at intervals on straight pipe runs sufficient for identification. Identification may be accomplished by stenciling, the use of tape, or markers. The legend should identify the contents of the piping system, indicate direction of flow, and provide sufficient additional details such as temperature, pressure, etc. as necessary to identify the hazard (see Figure 3.18). It is also a good practice to label valves that serve as key control points for blocking the flow of energy (see Figure 3.19).

Figure 3.19. Emergency shut-off valve with identification tag.

Hydraulic Energy

Hydraulic systems at a broad range of working pressures are widely used throughout industry. Machines and equipment that use hydraulic energy include brakes, presses, shears, extruders, hoists, and rams. Heavy-duty systems are capable of generating extremely high pressures of many tons.

Hydraulics, by definition, is the application of fluid mechanics to engineering devices involving liquids, usually water or oil. In a typical hydraulic system, a motor- or

Figure 3.20. (left) Hydraulically actuated molds on blowmolder.

Figure 3.21. (right) Hydraulic accumulator.

Figure 3.22. High-pressure water vessels.

engine-driven pump supplies high-pressure fluid through hoses, pipe, or tubing to a cylinder (see Figure 3.20). Force is transmitted through the fluid to drive a piston in the cylinder used to do work. Some hydraulic systems may employ a master cylinder that in turn can be used to drive one or more slave cylinders. The force applied to the master cylinder can also be multiplied in the slave cylinder (by using a larger piston in the slave).

In a simple hydraulic system, simply shutting off the electrical power to the hydraulic pump may be sufficient to accomplish energy control. Complex systems, however, may use a series of pumps, reservoirs, accumulators, cylinders, electrically operated valve blocks, manual and automatic valving, etc., and additional steps for hydraulic energy control may be needed. Hydraulic pressure, for example, can be stored in hydraulic lines, accumulators (as in Figure 3.21), and cylinders.

When the residual pressure is sufficient to move machine components or otherwise cause injury if released, the system needs to be isolated and the residual pressure relieved. Isolation of the system is accomplished by closing and locking/tagging out the appropriate control and section valves, and by capping or blanking hydraulic lines. Pressure relief may accomplished by opening the appropriate relief valves, and by moving machine components to a position where hydraulic energy is dissipated.

Hydraulic energy may also be found in water systems under pressure. Similar to systems using hydraulic oil as a pressure medium, pumps may also be used to push water at high pressure through piping systems and equipment. Energy isolation (by locking and tagging out the pump and line valves) and residual pressure relief must be provided when necessary to protect employees who service or maintain such high-pressure water equipment (see Figure 3.22).

A column of water, such as those found in municipal and industrial water supplies and drainage systems, may also exert moderate pressure in a piping system, often referred to as "pressure head." These systems may also need to be isolated and drained when necessary to prevent injury to maintenance and repair personnel.

Pneumatic Energy

Figure 3.23. Lockout air device. (Photo courtesy of Rockford Systems, Inc., Rockford, IL)

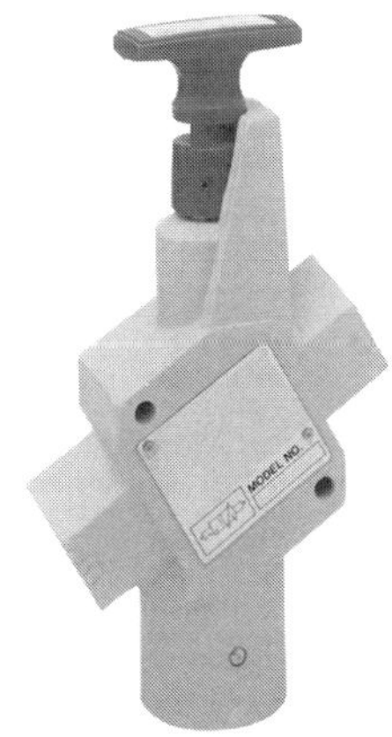

Pneumatic energy is encountered in pressurized systems containing mainly compressed air, and other gases to a lesser extent. Pneumatic systems include compressors, blowers, receivers, reservoirs, coolers, air dryers, controls, and distribution piping. Compressed air is used to power a great deal of industrial equipment. Plant air is often distributed throughout facilities for the operation of air-powered tools and equipment and other processes. Compressed air is also used in pneumatic instruments and controls, air motors, and purge connections. Blowers are used to provide air to heaters and other equipment.

Pneumatic lines and vessels can also retain energy in the form of stored pressure that may have to be relieved prior to service or maintenance. Isolation of the system is typically accomplished by shutdown and lockout/tagout of the compressor or blower, and closing and locking/tagging out the appropriate control and section valves (see Figure 3.23). Pressure relief may be accomplished by opening the appropriate bleed screws, relief valves, and vents. Capping or blanking of pneumatic lines may also be appropriate for high-pressure systems, and where line valves are not available or adequate for energy control.

Thermal Energy

Thermal energy is manifested by high or low temperature, and can result from mechanical work, radiation, combustion, chemical reaction, or electrical resistance. Contact with thermal energy can result in burns or frostbite; the severity of injury varies with the temperature and duration of exposure (see Figure 3.24). Thermal energy cannot be turned off in the conventional sense, and must be dissipated or controlled when necessary to prevent injury to service or maintenance personnel.

Figure 3.24. Extruder–400° F operating temperature.

Heat is the most common form of thermal energy found in the industrial environment, and is encountered with equipment such as heaters, burners, boilers, dryers, ovens, furnaces, heat exchangers, cupolas, extruders, hot water, and steam systems. Control of thermal (heat) energy can be accomplished in a number of ways depending on the complexity of the equipment or system. Often, thermal energy can be dissipated by simply allowing sufficient time after shutdown for parts to cool. Serious burns can result from contact with parts at temperatures above 140° F (60° C). As a general guideline, surface temperatures of exposed surfaces and parts should be allowed to cool to 110° F (43° C) or less where inadvertent contact might occur, or where the part or equipment will be handled.

If necessary, forced cooling methods using fans, water, and ice can be used to reduce cooling time. When allowing parts to cool is not feasible, insulating blankets, covers, gloves, or other heat-protective clothing and equipment may be needed to prevent contact with heat sources and allow the worker to perform the service or maintenance task safely. With a simple system such as an electric radiant heater, for example, shutting down the equipment (and locking/tagging out the local disconnect or main breaker if necessary) and allowing sufficient time for parts to cool is sufficient.

Steam and other hot gases can also be found in piping systems, and can pose a serious injury risk should unexpected release occur. Steam is typically generated by heaters (furnaces) and boilers (often combined in one unit, as in Figure 3.24), and is distributed through a system of piping, connections, valves, and fittings that are suitable for the pressures involved. Steam is used in many industrial processes, including petroleum refining and electrical generation (see Figure 3.25). In electrical generation, for example, steam leaves the boiler at the highest pressure required and is then is reduced in pressure in turbines that drive process pumps and compressors. Most steam used in the refinery is then condensed to water in various types of heat exchangers. The condensate is reused as boiler feedwater or discharged to a wastewater treatment system. When refinery steam is used to drive steam turbine generators to produce electricity, the steam must be produced at much higher pressure than required for process steam.

Thermal energy in piping systems can also be retained in the form of stored pressure that may have to be relieved prior to service or maintenance. Isolating the system is typically accomplished by first shutting down and locking/tagging out the equipment providing the heat source, including electrical and fuel sources when necessary.

The appropriate control and section valves must also be closed and locked/tagged out to isolate the section of piping system to be worked on. Pressure relief should be accomplished by opening the appropriate relief valves, vents, etc. Capping or blanking may also be appropriate where line valves are not available or as an additional measure. When dealing with high-risk sit-

Figure 3.24. Typical furnace/boiler installation.

Figure 3.25. Steam vessel.

uations such as superheated steam and supercooled or cryogenic liquids, the more protective isolating methods such as the double block and bleed arrangement should be used to isolate and remove residual energy from the piping system.

Chemical Energy

Chemical energy is associated with chemical reactions, processes, and systems. During chemical reactions, reactants are converted or synthesized into other compounds, or may decompose into other products. Common types of reactions include oxidation-reduction, neutralization, and hydrolysis. Depending on the types of substances involved, chemical reactions can also generate heat or pressure (for example, during exothermic reactions or polymerization), or evolve toxic or flammable fumes or gases. Some materials may react violently with others (for example, water-reactive materials such as sodium metal), or pose a risk of fire or explosion if the reaction is uncontrolled. Chemical energy can also be converted to electrical energy, as in the case of batteries and galvanic cells.

Safe design and construction of chemical systems, along with special operating procedures, usually ensures the safe operation and control of chemical processes during production. The control of chemical process hazards during service or maintenance tasks, however, may pose unique challenges (see Figure 3.26). Isolation of chemical processes and systems will vary depending on the complexity of the equipment and materials involved. Generally, conventional means (as used with other piping systems) are used to isolate the system by shutdown and lockout/tagout of the chemical feed pumps, mixers, and other associated equipment, along with closing and locking/tagging out the appropriate control and section valves. In addition to shutting down pumps and turning off valves, control devices such as level controls and pressure controllers may also have to be safely positioned by following special procedures to deactivate the process (see Figure 3.27). In some cases, a waiting period may be necessary to allow a reac-

Figure 3.26. Complex chemical systems may pose unique isolation challenges. (Photo courtesy of Smith Flow Control [USA] Division of Castell Interlocks Inc.)

Figure 3.27. Level controls may also have to be safely positioned to control chemical energy.

tion to proceed to completion. A predetermined sequence of steps may be needed to ensure that process shutdown is accomplished safely.

Figure 3.28. Removal of chemical piping system contents may be needed to prevent unexpected release of stored chemical energy.

The presence of residual chemicals in piping systems and vessels may also pose risks to service or maintenance personnel due to contact with hazardous substances or the unexpected release of stored chemical energy. Buildup of heat or pressure, for example, could result if materials in the system were incompletely reacted, or if reactive residual chemicals were to otherwise come into contact. Pressure relief or removal of piping system contents may therefore be needed, and may be accomplished by opening the appropriate bleed, drain vent valves, flares, etc. Decontamination and the purging and flushing of lines and vessels may also be required (see Figure 3.28). Caution should be exercised, however, to ensure that toxic, flammable, or explosive substances are not vented directly to the atmosphere, and that hazardous materials and wastes are properly collected, stored, and disposed of so they do not pose health or environmental risks.

Although isolating chemical piping systems is primarily accomplished using valves, capping or blanking may also be used effectively where line valves are not available or adequate for energy control. When dealing with toxic, flammable, corrosive, and reactive substances, the more protective isolating methods such as the double block and bleed arrangement should be used to isolate and remove residual energy from the piping system. The double block and bleed is accomplished by closing and locking/tagging two in-line valves (upstream toward the source) and by opening and locking or tagging a drain or vent valve in the line between the two closed valves. A variation of this method includes using a blank or blanks in place of the valves.

In some cases, a small amount of leakage or drainage of substances can occur and is unavoidable, such as with supercooled or cryogenic liquids, even with the system isolated. A certain amount of leakage may be acceptable if it occurs without endangering employees, but precautions such as continuous monitoring may be needed to ensure that such leakage does not approach a hazardous level (for instance, air monitoring to ensure that hazardous levels of toxics, flammables, or oxygen deficiency do not occur). In addition, appropriate personal protective equipment should be worn whenever there is a potential for chemical exposure.

Alternative Measures

The lockout/tagout standard requires that machines, equipment, and processes be de-energized and locked/tagged out before performing any maintenance, where unexpected start up or release of hazardous energy could occur and cause injury. In a production environment where conveyors, labelers, fillers, packers, palletizers, and other types of equipment are found, unjamming, lubricating, cleaning, adjusting, and other minor servicing tasks can be frequent and necessary. Having to perform lockout on a machine each time a minor servicing task is performed is often time consuming and can adversely affect production. OSHA understood this potential impact when it promulgated the lockout/tagout standard, and included a "minor servicing exception" to provide employers with flexibility in performing such tasks. The exception states:

> Minor tool changes and adjustments, and other minor servicing activities, which take place during normal production operations, are not covered by this standard if they are routine, repetitive, and integral to the use of the equipment for production, provided that the work is performed using alternative measures which provide effective protection.

Many production managers would like to use this exception to maintain or improve productivity; the key to using it is to eliminate maintenance task hazards by using machine guarding or other methods that provide effective protection. The following discussion is intended to provide further guidance on the applicability of the minor servicing exception, and recommendations for how excepted tasks can safely be done through the use of specific control devices, work practices, employee training, and other measures.

To take advantage of the minor servicing exception, four criteria must be met. First, the servicing activity must be minor. Generally, servicing is not considered to be minor in nature if it involves extensive disassembly of the machinery or equipment, removal of any major parts of the equipment such as panels or other barriers that restrict access to moving mechanical parts or energized electrical equipment, or in some cases, if it involves more than a single person to perform the operation.

Second, the servicing activity must take place during normal production operations, while the machine is actually performing its intended production function. Generally, anything that is done to prepare a machine to perform its normal production operation, such as changing a machine part or setting up, is not considered utilization of a machine and should be classified as servicing or maintenance rather than normal production operations.

Third, the servicing activity must be routine, repetitive, and integral to the use of the machine for production.

Fourth, the servicing activity must be performed in a manner that protects employees from hazards by effective alternative means. Activities that do not meet all of these elements do not qualify for the exception, and must be performed with the machine de-energized and all potentially hazardous energy locked/tagged out.

To be considered effective, the alternative protective measures must enable the employees to safely perform the servicing task without being exposed to the unexpected activation of the machine or the release of stored energy. Acceptable alternative measures may include the use of specially designed tools, control devices, work practices, employee training, and other measures. One or more of these measures may be needed to provide effective protection. Some examples include:

- The use of specially designed servicing tools or remote devices that allow the employee to perform the task without placing any part of his or her body in a danger zone (caution should be exercised to ensure that the tool itself does not create a hazard).
- The use of local disconnects that are under the exclusive control of the employee performing the minor servicing. Effective alternative protection may not include, by themselves, simple pushbuttons, selector switches, and other control circuit-type devices.

Figure 3.29. Door interlock on Haas mill utilizes highly reliable safety-rated components that allow safe access to the point of operation for performing minor servicing.

- The use of machine safeguarding that protects employees throughout service tasks from machine- or equipment-related hazards, such as those encountered at the point of operation, power transmission apparatus, nip points, and other moving parts. Safeguarding options include fixed guards or enclosures, barrier devices, interlocked barrier guards, and other electronic safety devices such as light curtains and other presence-sensing devices. Simply shutting off the machine, however, is not a recognized machine safeguarding method. It should also be noted that OSHA regulations for the selection and use of work practices [29 CFR 1910.333(b)(2)(ii)(B)] pertaining to electrical work, prohibit the use of interlocks for electric equipment as a substitute for lockout/tagout procedures.

Where safeguarding methods are used as the alternative protective measures, employees must not remove or bypass guards or otherwise place any part of their body into an area where they may be subject to injury from unexpected energization or activation of the machine. Other measures may be needed in addition to the safeguarding to prevent unexpected energization or activation of the machine due to the release of stored or residual energy, or from human error. With respect to the particular safeguarding methods and devices employed, it is a good idea to conduct hazard analyses to demonstrate that effective protection is provided.

Several factors should be considered in evaluating the effectiveness of safeguarding methods and devices such as interlocks. For example, safeguards should:

- Comply with applicable OSHA, ANSI, or other standards;
- Be designed, constructed, and installed so that they cannot readily be removed, bypassed, overridden, or otherwise defeated; and
- Incorporate highly reliable safety-rated components and circuitry (control-reliable systems) to ensure that they will not malfunction or fail (for example, due to failure of a spring, sticking or welding of contacts, or shorting across electrical circuits) and cause injury (see Figure 3.29). The equipment manufacturer may need to be consulted in this regard.

In addition, safeguards should be inspected frequently (for example, tested before each shift) to ensure proper operation.

Careful consideration should be given to the effectiveness of alternative measures in using the minor servicing exception in lieu of lockout/tagout. In those cases where de-energizing and locking/tagging out are not feasible for valid reasons, only alternative procedures that meet rigid safety criteria should be used. Some sage advice in this regard, is "when in doubt, lock it out." In addition, where the exception is used, written procedures (standard work practice instructions) should be prepared that describe the work methods to be used and any needed precautions, and employees should be trained in their use.

Other examples of alternative methods of isolating or blocking energy and securing the points of control are also provided in Table 3.1, taken from the NIOSH publication *Guidelines for Controlling Hazardous Energy During Maintenance and Servicing*.

Table 3.1: Examples of alternate methods of isolating or blocking energy and securing the points of control.

Method of isolating or blocking energy	Method of securing the point of control	Remarks
Mechanical motion (rotation, translation, linear, oscillation)		
1. Remove segments of operating mechanical linkages such as dismantling push rods or removing belts, flywheels.	• Tag the linkages and place them in a locked cabinet away from the machine, or • Attach warning tags where the linkages were removed and restrict access to trained personnel, or • Post a person to protect against unauthorized reinstallation of the linkages.	
2. Use blocking devices such as wood or metal blocks.	• Chain and lock in point of control or use metal pins driven or welded in place, or • Attach warning tags on the blocking devices and restrict access into the area to trained personnel, or • Post a person to protect against unauthorized removal of the blocking devices.	
3. Remove power or energy from the driving mechanism, such as a) disconnect main electrical source, or b) close hydraulic or pneumatic valves, bleed.	• Padlock in the Off position then disconnect pneumatic and hydraulic lines and tags, or • Attach warning tags at control points and restrict access to trained personnel, or • Post a person to protect against unauthorized reinstallation of the linkages.	Check for alternate sources of power. Check for residual pneumatic and hydraulic energy.

Table 3.1: Examples of alternate methods of isolating or blocking energy and securing the points of control. (Continued)

Method of isolating or blocking energy	Method of securing the point of control	Remarks
Electrical		
1. Place the main electrical disconnect switch in the Off position.	• Secure by a padlock, a clip and padlock, or a bar and padlock, or • Attach warning tag and restrict access into the area to trained personnel, or • Post a person to protect against unauthorized actuation of the switch.	Check for alternate sources or power.
2. Remove segments of electrical circuit, such as printed circuit modules.	• Tag the module and place in a locked cabinet away from the control center and tag the control center door, or • Attach warning tag at the module location and restrict access to trained personnel, or • Have a person remain at the control center to protect from unauthorized installation of a spare or replacement module.	
Thermal (steam)		
Close valves and maintain an open bleed.	• Chain and padlock valve or use blind flanges or slip blinds, or • Attach warning tags to the valves and restrict access to trained personnel, or • Station a person at the valve location to protect against unauthorized or inadvertent opening of valves.	Allow time for residual heat to dissipate.
Potential (pressure)		
Close valves and maintain open vent to relieve.	• Secure, block, blind flange, slip blind, or valve with locking device, or • Attach warning tags and restrict access to the area to trained personnel, or • Station a person at the valves to protect against unauthorized actuation.	
Potential (gravity)		
Block in place by using metal or wood blocks under the mechanism or pin the linkages in a position where gravity will not cause the mechanism to inadvertently fall.	• Secure, block, or pin with a locking device, or • Attach warning tags to blocks, linkages, and pins and restrict access to trained personnel, or • Station a person at the mechanism to prevent unauthorized removal of blocks and pins and reinstallation of linkages.	Energy could be dissipated by lowering to a point where gravity could no longer cause inadvertent falling.

Table 3.1: Examples of alternate methods of isolating or blocking energy and securing the points of control. (Continued)

Method of isolating or blocking energy	Method of securing the point of control	Remarks
Potential (springs)		
Block in a safe position by pinning or clamping the device, eliminating the potential or unrestricted and undesired travel.	• Secure, pin, or clamp in place with a locking device, or • Attach warning tags to the pins and clamps and restrict release or access to trained personnel, or • Station a person at the control point to protect against pin or clamp removal and unauthorized activation of the spring mechanism.	Spring energy could be dissipated by releasing or dismantling the mechanism.

Chapter 4

Lockout/Tagout Preparation, Application, and Removal

OSHA's lockout/tagout standard does not prescribe the specifics of the energy control measures to be employed. Rather, the standard stipulates the general procedures and performance to be achieved, giving employers flexibility in choosing the control procedures to be implemented. (Figure 4.1 presents a functional flow diagram for implementing lockout/tagout.) For example, the standard sets forth nine general principles which provide the framework for ensuring the safety of personnel who maintain machines (see 1910.147(c)—*General*):

Energy Control Program. Employers must establish a program that includes energy control procedures and employee training to ensure that, before any employee performs maintenance on a machine where the unexpected energizing, start up, or release of stored energy could occur and cause injury, the machine must be isolated and rendered inoperative. Establishing an energy control program is the basic tenet of the standard.

Lockout/Tagout. Lockout (as opposed to tagout) should be used when the machine's energy isolating devices are capable of being locked out. Tagout, without lockout, may be used only when the energy isolating device is *not* capable of being locked out or when the employer can demonstrate that use of the tagout system alone will provide "full employee protection" (see below). Isolating devices should be capable of being locked out in the majority of circumstances.

Full Employee Protection. When a tagout system alone is used, it must provide a level of safety equivalent to that obtained by using lockout, and the tagout device must be attached at the same location the lockout device would have been attached. Additional protective measures are needed to demonstrate that full employee protection has been achieved (more on this later). In addition, if an energy isolating device is not capable of being locked out, it must be redesigned (retrofitted or replaced) to accept a lock whenever major replacement, repair, renovation, or modification of the machine is performed.

Energy Control Procedures. Written machine-, equipment-, and process-specific procedures must be developed and used to protect employees who perform maintenance on machines.

Protective Materials and Hardware. Devices such as locks, tags, chains, key blocks, adapter pins, wedges, self-locking fasteners, and other hardware must be provided to employees for isolating, securing, or blocking machines from hazardous energy sources.

Figure 4.1. Functional flow diagram for implementation of lockout/tagout requirements. (Source: OSHA.)

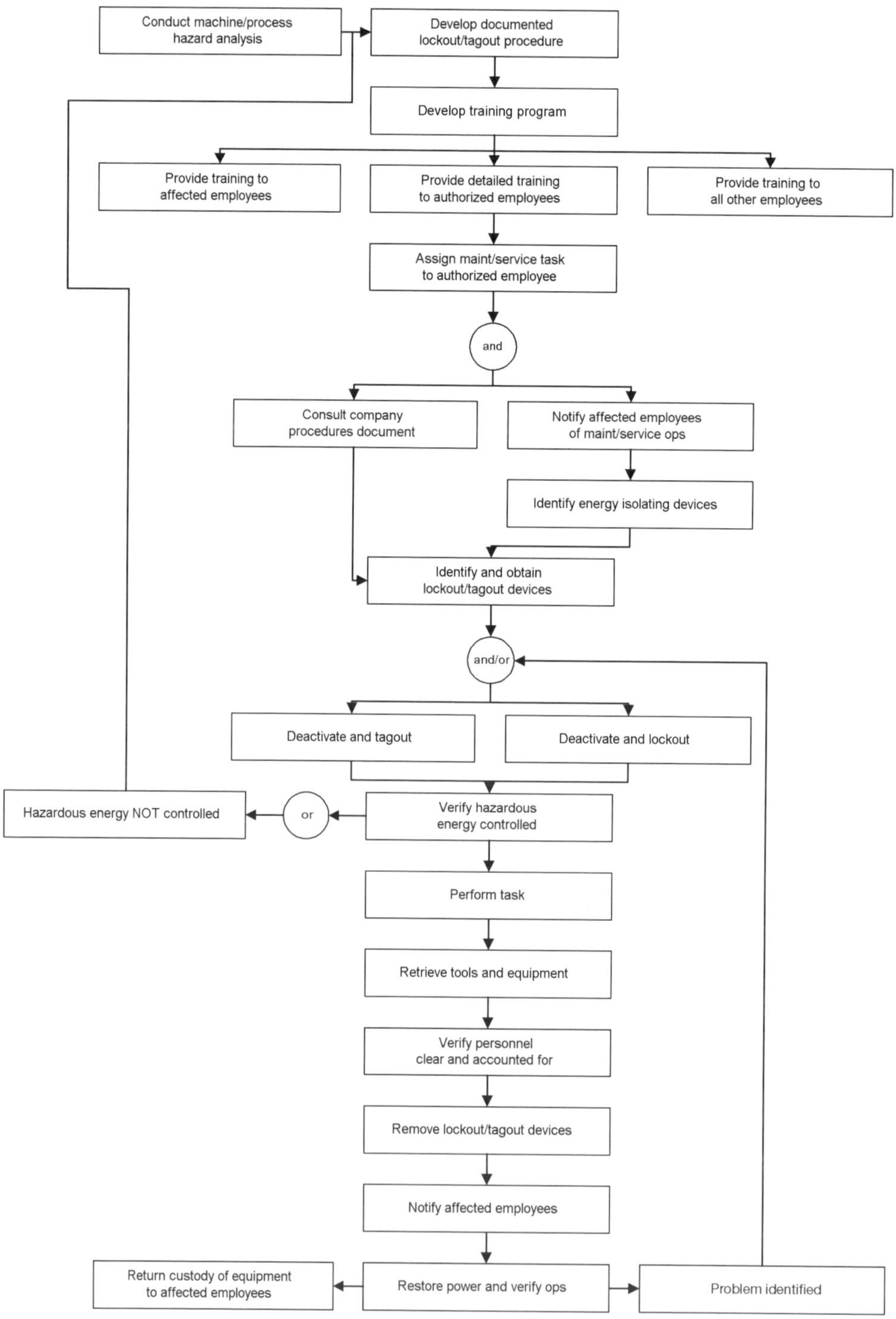

Periodic Inspection. To ensure that energy control procedures are adequate and are being followed, inspection of the procedures must be conducted periodically (at least annually) and documented. The inspection must be performed by a person other than the one using the energy control procedure being inspected, and must include a review between the inspector and each authorized employee who has responsibilities under the procedure being inspected. Where a tagout system alone is used, the review must also include affected employees. Any deviations or inadequacies identified during the inspection must be corrected.

Training and Communication. Training must be provided to all authorized and affected employees prior to assignment to ensure that they understand the purpose and function of the energy control program and have the knowledge and skills required for the safe application, use, and removal of energy controls.

Energy Isolation by Authorized Employees Only. Implementation of lockout and tagout system procedures may be performed by authorized employees only. An authorized employee is an employee who locks out or tags out a machine to perform maintenance on it. Authorized employees have the requisite training and knowledge to recognize the types and magnitude of the hazardous energy sources present, and the methods and means necessary for energy isolation and control.

Notification of Affected Employees. Affected employees must be notified (for instance, by the authorized employee) of the application and removal of lockout and tagout devices. An affected employee is an employee whose job requires him or her to use a machine on which servicing is being performed under lockout or tagout, or whose job requires him or her to work in an area in which such maintenance is being performed. Notification must be given before the controls are applied, and after they are removed from the machine.

Sequence of Lockout/Tagout

In addition to these nine general principles, Section (d) of the standard (*Application of Control*) prescribes six specific elements and actions for implementing lockout/tagout system procedures, *and specifies the required sequence in which they must be done* (see Figure 4.2):

1. **Preparation for Shutdown.** Before an authorized or affected employee turns off a machine, the authorized employee must have knowledge of (must determine) the types and magnitudes of the energy, the hazards of the energy to be controlled, and the method or means to control the energy. This first step is intended to allow employees to get ready for the shutdown, and plan ahead to safely perform the intended maintenance task. In complex or group lockout/tagout situations, all involved personnel should be made aware of their respective roles and assignments.

2. **Machine or Equipment Shutdown.** To place the machine in a state that will permit employees to work on it safely, it must be turned off or shut down using established procedures. An orderly shutdown must be used to avoid any increased or additional hazards to employees as a result of the shutdown. Typically, the machine is shut down by the "normal stopping means," or by following established run down procedures. Commonly, shut-

Figure 4.2. Functional flow diagram for accomplishing energy isolation. (Courtesy of Office of Environmental, Health and Safety, Wayne State University. Detroit, Michigan.)

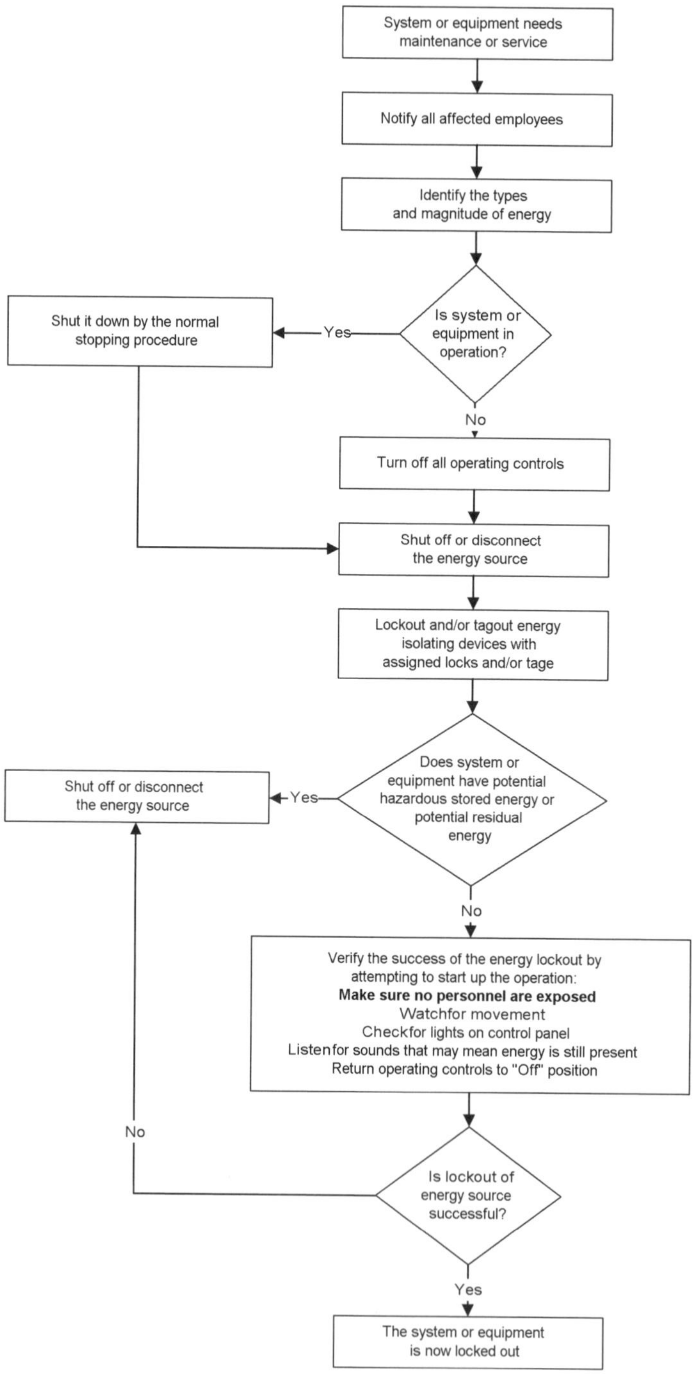

down is accomplished with the machine operating controls (for instance, by depressing the "Start/Stop" button or "Power On/Off" switch, or opening a toggle switch). Complex equipment, however, may require shutdown of multiple machine components. With computer operated machines and equipment, remote shutdown at a computer workstation or programmable logic controller (PLC) may also be required.

3. **Machine Isolation.** All energy isolating devices that are needed to control the energy to the machine must be physically located and operated so that the machine is isolated from its energy sources. Generally speaking, isolating devices should be placed in the off, or safe position; electrical isolating devices such as breakers, disconnects, and knife switches are opened. Valves that control hydraulic, pneumatic, chemical, or thermal (for instance, steam) systems are closed. If necessary, mechanical power transmission apparatus such as chains, belts, shafts, gears, etc. may be disconnected or removed.

4. **Lockout/Tagout Device Application.** Isolating devices must next be locked or tagged out by the authorized employees using individual locks and tags. Locks and tags must be affixed to each energy isolating device to hold it in a safe or off position and physically prevent its operation. Tagout devices (tags) must be affixed in a manner that indicates movement of the isolating device from the safe or off position is prohibited. Where machines are capable of being locked out, tags should be affixed to the isolating devices along with locks by placing the shank of the lock through the grommet provided on the tag. Where use of a tagout system alone is permitted and used, the tag must be fastened at the same point at which the lock would have been attached. When a tag cannot be affixed directly to the isolating device, it must be placed as close as safely possible to the device in a position that is immediately obvious to anyone who may attempt to operate the device.

5. **Stored Energy Release/Restraint.** After locks and tags are applied to the energy isolating devices, the next prescribed step is for all potentially hazardous stored or residual energy to be relieved, disconnected, restrained, or otherwise rendered safe. Although hazardous energy has been isolated in previous steps, stored, residual, or potential energy may still be retained in the machine, equipment, or process. For example, potentially hazardous stored energy may be found in springs that are under tension or compression, or in raised loads that could fall if dislodged (potential energy due to gravity). To control the potentially hazardous energy, springs should be brought to a relaxed or neutral position, and components that could move and cause injury should be moved to a stable, resting position or blocked/restrained. Electrical energy that might endanger personnel can also be stored in capacitors. Capacitors and other high capacitance circuit elements may need to be discharged, short-circuited, or grounded to be rendered safe. Residual hydraulic or pneumatic energy may be trapped in vessels, lines, and piping systems downstream from isolating valves and may need to be bled down and blocked/blanked. Chemical processes may also harbor chemical energy in vessels or piping systems. The contents of chemical piping and processing systems may need to be removed and the system decontaminated to deactivate the process.

6. **Verification.** Prior to starting maintenance work on machines that have been locked out and/or tagged out, the authorized employee must verify that isolation and de-energization of the machine has been accomplished. Machines must be checked with test instruments

and/or by visual inspection to verify that energy isolation has been accomplished. An attempt to restart the machine at the operating controls should also be performed where possible to verify that it cannot be restarted. If there is a possibility for reaccumulation of stored energy to a hazardous level, verification of isolation must continue until the maintenance is completed, or until the possibility of such accumulation no longer exists.

Release From Lockout/Tagout

After the maintenance task is complete, the machine will be reenergized and restored to normal production operations. Before the machine is restarted and returned to service, OSHA's lockout/tagout standard also requires that standardized procedures be followed and actions taken to ensure that employees are not subject to an increased potential for injury (see Figure 4.3). The three prescribed procedures or actions for release from lockout/tagout are described in section (e) of 1910.147:

1910.147 (e)

(1) *The machine or equipment.* The work area shall be inspected to ensure that nonessential items have been removed and to ensure that the machine or equipment components are operationally intact.

(2) *Employees.* (i) The work area shall be checked to ensure that all employees have been safely positioned or removed.

(ii) Before lockout or tagout devices are removed and before machines or equipment are energized, affected employees shall be notified that the lockout or tagout devices have been removed.

(3) *Lockout/Tagout Device Removal.* Each lockout or tagout device shall be removed from each energy isolating device by the employee who applied the device. *Exception to paragraph (e)(3):* When the authorized employee who applied the lockout or tagout device is not available to remove it, that device may be removed under the direction of the employer, provided that specific procedures and training for such removal have been developed, documented and incorporated into the employer's energy control program. The employer shall demonstrate that the specific procedure provides equivalent safety to the removal of the device by the authorized employee who applied it. The specific procedure shall include at least the following elements:

(e)(3)(i)
Verification by the employer that the authorized employee who applied the device is not at the facility;

(e)(3)(ii)
Making all reasonable efforts to contact the authorized employee to inform him/her that his/her lockout or tagout device has been removed; and

(e)(3)(iii)
Ensuring that the authorized employee has this knowledge before he/she resumes work at that facility.

Paragraph (e)(1) requires that when the maintenance task is complete, the work area be inspected to ensure that nonessential items have been removed, and that the equipment is operationally intact and in a condition to be safely reenergized. This inspection is needed since accidents can still occur after the maintenance task is complete. Missing guards and errantly placed or left behind tools, parts, and materials, for example, can present injury risks to personnel, or result in equipment damage upon startup. Depending on the complexity of the equipment and the type maintenance performed, visual inspection may be adequate, or it may need to be supplemented with checklists and other administrative procedures.

Paragraph (e)(2) then requires a check of the work area to ensure that all employees are safely positioned, and that affected employees are notified that lockout/tagout devices have been removed and that the equipment is to be reenergized. A simple procedure to follow to verify that the work area and machinery is ready to be restarted is for a supervisor, foreman, or other lead person to ask the maintenance workers if they are done, and then visually inspect the area. Warning devices such as bells, horns, buzzers, or strobe lights are also useful for alerting employees to stand clear.

Paragraph (e)(3) requires that a lockout/tagout device be removed from an energy isolating device *only* by the employee who applied the device. This general rule is intended to provide assurance to authorized employees that the lockout/tagout devices will remain on the isolating devices and protect them from energy hazards for the duration of the service or maintenance task.

An exception to this general rule is noted in (e)(3) that permits removal of an employee's lockout or tagout device by another employee when the employer can demonstrate a need to do so, such as *during an emergency.* OSHA emphasizes, however, that removal of a personal lockout or tagout device by another person may not be based on convenience or simple unavailability of the employee. In addition, specific alternative procedures for such removal must be developed and used, and employees must be trained in their implementation. The procedures must include, at minimum: 1) verification by the employer that the authorized employee who applied the device is not at the facility; 2) making all reasonable efforts to contact the authorized employee to inform him or her that his or her lockout or tagout device has been removed; and 3) ensuring that the authorized employee knows that his or her lockout or tagout device has been removed before he or she resumes work at that facility.

To implement the requirements of paragraph (e)(3), the "one lock, one key rule" is typically applied. Under this rule, locks are singularly keyed, and individual keys are retained by and kept under the exclusive control of the authorized employee. The unavailability of duplicate keys to other persons helps prevent the unauthorized removal of locks. In consideration of the noted exception to (e)(3), duplicate keys for use by supervisory personnel during emergencies are permitted; it is critical, however, that access to keys be controlled and secure.

The above steps for de-energizing and locking/tagging out machines, equipment, and processes are only a part of the total energy control program that must be implemented. In addition, employees must have written procedures for all covered machines and must be trained in their duties and responsibilities under the energy control program. Periodic inspections must also be conducted to ensure the continued effectiveness of the program.

Figure 4.3. Functional flow diagram for returning equipment to service. (Courtesy of Environmental, Health and Safety, Wayne State University, Detroit, Michigan.)

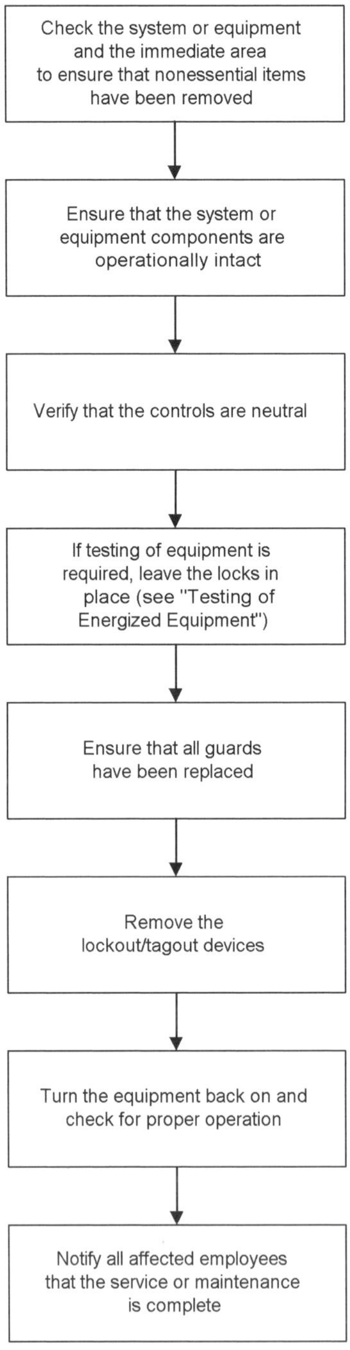

Chapter 5

Verification of Energy Control

The verification of energy control is fundamental to assuring the safe performance of service or maintenance on machines, equipment or processes. After all hazardous energy is isolated or controlled, and following the application of locks and/or tags, it is critical to verify that de-energization and isolation have been effectively accomplished before proceeding with service or maintenance. Verification of isolation must be continued until the servicing or maintenance is completed, or until the possibility of reaccumulation of stored energy no longer exists. The following case studies illustrate just how critical verification is.

Verification Case Study 1

A millwright was assigned to perform a maintenance task on an overhead crane. Following the company's prescribed energy control procedure, he opened the crane's electrical disconnect switch and placed his lock onto this energy isolating device. The disconnect switch lever, however, was corroded, allowing it to physically separate from the disconnect switch assembly. As a result, when the millwright placed the lock on the lever in the "off" position, the stub located inside the electrical box remained in the "on" position. The employee did not attempt to verify de-energization by using the crane start button, or testing for no voltage. When he contacted the energized electrical bus bar located near the bridge railings, he was electrocuted.

Verification Case Study 2

A lineman was assigned to troubleshoot a fault in a power distribution system. To perform this task, he entered a manhole that provided access to an underground raceway that contained several energized 600-volt cables. After identifying the cable with the fault, he asked his assistant at street level to hand him a tag, which he attached to identify the problem cable. He then exited the manhole to retrieve the equipment needed to cut and pull the cable so it could be replaced. After confirming de-energization of the problem cable with the power dispatcher, he reentered the manhole. He then attached cable cutters to the tagged cable and proceeded to cut through a live, 600-volt cable. The man was permanently disabled as a result of burns caused by the resulting electrical arc. He had tagged and cut through the wrong cable! The simple step of verifying the de-energization of the cable he was about to cut would have prevented this accident.

Paragraph (d)(6) of the lockout/tagout standard requires that the authorized employees verify that isolation and de-energization have been effectively accomplished before starting servicing/maintenance work. This action is intended to assure the employees that the machine or equipment is isolated from the energy, that residual or stored energy has been dissipated or blocked, and that injury could not result from the inadvertent activation of the operating controls. Workers must not proceed with service or maintenance unless de-energization and isolation of the machine, equipment, or process have been verified.

> 1910.147 (d)(6)
>
> *Verification of isolation.* Prior to starting work on machines or equipment that have been locked out or tagged out, the authorized employee shall verify that isolation and deenergization of the machine or equipment have been accomplished.

Verification is the sixth step in the energy control procedure sequence prescribed in Section (d) of the standard (Application of Control), and follows preparation for shutdown, shutdown, machine or equipment isolation, lockout/tagout device application, and stored energy relief/restraint. Verification is done before the servicing or maintenance work begins to ensure that these previous steps have been accomplished. For example, the authorized employee needs to verify that: the machine or equipment has been turned off or shut down properly as required by paragraph (d)(2) of this standard; that all energy isolating devices were identified, located, and operated as required by paragraph (d)(3); that the lockout or tagout devices have been attached to energy isolating devices as required by paragraph (d)(4); and that stored energy has been rendered safe as required by paragraph (d)(5). If there is a possibility for reaccumulation

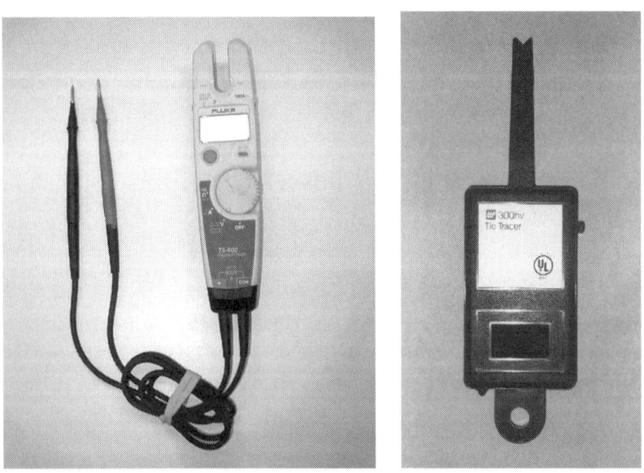

Figure 5.1. Verification tools (clockwise from top left): Electrical tester (multimeter); Tic-Tracer; AC sensor; pressure gauge; flow transmitter; temperature gauge.

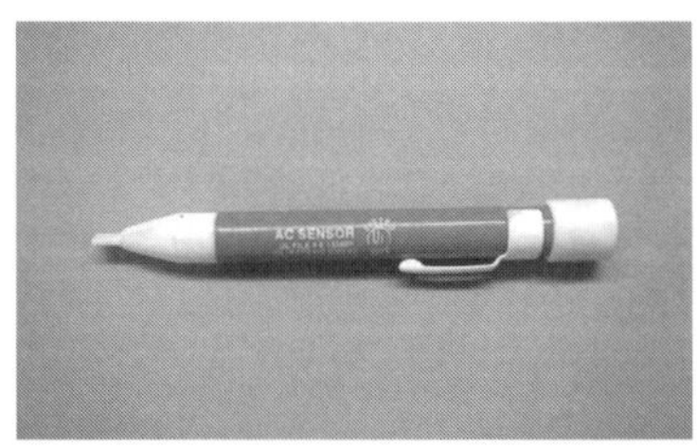

of stored energy to a hazardous level, verification of isolation must continue until the service or maintenance is completed, or until the possibility of such accumulation no longer exists.

Verification of isolation may involve a deliberate attempt to start the equipment that has been isolated, the use of appropriate test instruments or equipment, or visual inspection. In some cases, a combination of these verification methods may be needed. The appropriate method or combination of the methods will depend upon the forms of energy encountered, the type of machinery or equipment involved, and other factors. With most industrial machines and equipment, verification should be accomplished by attempting to restart the equipment at the operating controls, and confirming that the equipment cannot be restarted. As a precaution, the restart attempt should only be done after ensuring that no one is exposed. In addition, the operating controls should be returned to a safe position (for example, "neutral" or "off") to ensure a controlled startup after service or maintenance is completed.

Test Instruments and Equipment

Another way to verify the de-energization of machines or equipment is through the use of appropriate test instruments and equipment. Various types of instruments and equipment are available to test for or indicate the presence of hazardous energy (or likewise, to confirm that hazardous energy is not present), including portable, hand-held instruments such as electrical testers, gas monitors, and temperature probes; and fixed equipment such as pressure or temperature gauges, flow measuring devices, and so on (see Figure 5.1). Some of these devices use mechanical, dial-type indicators, while others are electronic. Gauges and flow indicators, for example, are often strategically placed at various points in a piping system to indicate pressure or flow. Complex equipment or processes may employ sensing, controlling, and data acquisition (SCADA) systems that gather information about pressure, flow, temperature, etc. remotely, and corroboration of de-energization and isolation at a PLC (programmable logic controller) or workstation may also be needed. When using gauges and other instrumentation for verification purposes, however, always exercise caution. Gauges and instruments can fail and give false readings. When possible, therefore, don't rely solely on these devices for verification, and use other available verification methods such as the equipment restart attempt and visual inspection.

Verification of Electrical De-energization

In cases involving electrical circuits and equipment, an electrical voltmeter or multimeter may be used to determine that there is no electrical energy available to the machine, or that conductors to which workers may be exposed are dead. Verification tests may also be needed to determine if any energized condition exists (or develops) as a result of inadvertently induced voltage or unrelated voltage backfeed,

Figure 5.2. Use of a digital multi-meter to confirm de-energization of electrical components on an uninterruptable power supply.

even though specific parts of the circuit have been de-energized and presumed to be safe (see Figure 5.2).

When working with electrical test instruments and equipment, it is important to follow the provisions of OSHA's Electrical Safety-Related Work Practices standard (29 CFR 1910.333). In this regard, only qualified employees who have been trained to work safely with test instruments and equipment on energized circuits are permitted to perform electrical testing work where there is a danger of injury due to accidental contact with energized parts or improper use of the test instruments or equipment.

> 1910.333 (c)—*Test Instruments and Equipment*
>
> (1) *Use.* Only qualified persons may perform testing work on electric circuits or equipment.
>
> (2) *Visual Inspection.* Test instruments and equipment and all associated test leads, cables, power cords, probes, and connectors will be visually inspected for external defects and damage before the equipment is used. If there is a defect or evidence of damage that might expose an employee to injury, the defective or damaged item will be removed from service, and no employee may use it until necessary repairs and tests to render the equipment safe have been made.
>
> (3) *Rating of Equipment.* Test instruments and equipment and their accessories will be rated for the circuits and equipment to which they will be connected and will be designed for the environment in which they will be used.

As a precaution, electrical test instruments and equipment should always be checked for proper operation on a known, live electrical circuit immediately before verification use. In the case of high voltage (greater than 600 volts), test instruments and equipment must also be tested immediately after use.

Verification of De-energization in Piping Systems

Piping systems that contain chemicals, steam, water, hydraulic fluid, compressed air, and other gases have the potential to cause serious injury or harm to workers if the hazardous energy within them is released unexpectedly. Potential energy is often encountered in these systems in the form of stored or trapped pressure, vacuum, or residual chemicals that could cause thermal or pressure buildup, and must be properly relieved, dissipated, or otherwise controlled when necessary. Verification of de-energization and isolation assures the service or maintenance worker that the system is in a condition to be safely worked on.

With piping systems, verification should ensure that the appropriate control and/or section valves have been closed and locked/tagged, or that the system has been isolated by other means such as blanking or blinding, or misaligning or removing sections of lines, pipes, or ducts. If it is necessary to vent, drain, or bleed the system to render it safe, involved personnel should ensure that the appropriate drain or relief valves have been opened, and verify that energy in the system has been reduced to a safe level. When possible, the completion of draining or venting should be verified by checking available pressure or level gauges. In pressurized systems, for example, verification can generally be accomplished by ensuring that the appropriate gauge indicates zero pressure in the part of the system to be worked on. Similarly, temperature gauges can be used to verify that thermal energy has been reduced to a safe level (see Figure 5.3).

Figure 5.3. Portable instruments are available to monitor high and low temperatures and pressure with optional probes. (Photo courtesy of Lumidor Safety Products, Miramar, FL.)

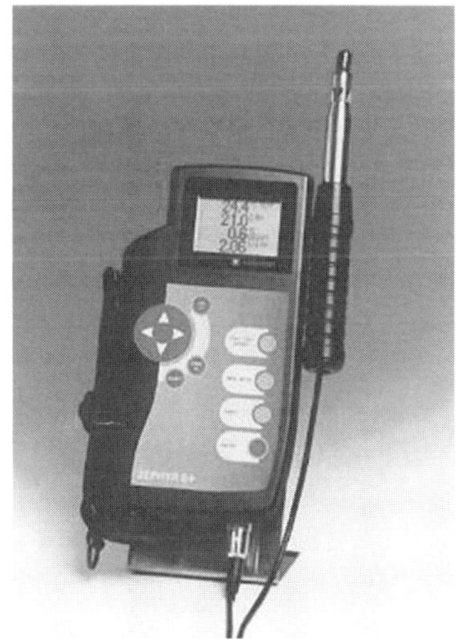

Where hazardous energy may be present in a piping system, and draining, venting, or pressure relief cannot be confirmed (for example, due to a lack of available gauges, indicators, or instrumentation), alternative measures (such as line breaking or pipe opening procedures) that provide effective protection must be used.

Visual Inspection

Visual inspection procedures are also of critical importance in verifying the isolation of machines and equipment. Before proceeding with service or maintenance, visual inspection should be used to confirm that operating controls are safely in the "off" position and that isolating devices have been properly secured in the "off" or "safe" position. For example, observing the position of electrical isolating devices such as breakers, disconnects, and knife switches can confirm that they are in the open or "off" position. At the same time, workers can visually verify that locks and/or tags have been applied to the isolating devices to secure them against movement.

Visual inspection should also be used to ensure that moving parts such as flywheels, blades, gears, and shafts have come to a complete stop, that raised or suspended parts have been lowered, and that any needed blocks, pins, bars, chains, or other devices are in place to secure potentially dangerous components against movement. The disconnection or removal of mechanical power transmission apparatus such as chains, belts, shafts, and gears should also be visually verified when such action is needed to prevent hazardous machine movement.

With piping systems, visual inspection can verify that the appropriate sectioning and/or control valves have been closed, that blanks/blinds are in place, and that drain, pressure relief, or vent valves are opened when needed. With chemical processes, it may also be necessary to visually verify the safe positioning of control devices such as level controls and pressure controllers, or the removal of system contents. Where it is necessary to drain residual chemicals in piping systems, it may be possible to visually verify that the contents have been removed by means of a level indicator, sight tube, or observation through a porthole or opening (following appropriate precautions).

Verification Continuance

Section (d)(5)(ii) of the standard requires that verification of energy isolation be monitored as frequently as necessary if there is a possibility of reaccumulation of stored energy. Where frequent or continuous verification of energy isolation is needed, monitoring may be accomplished in a number of ways, including visual observation or use of a monitoring device that will sound an alarm if a hazardous energy level is being approached.

Some leakage can occur and may be unavoidable in piping systems that hold energy-containing substances (cryogenic liquids, flammable gases, etc.). A certain amount of leakage may occur after isolation, and may be acceptable if workers are not endangered. However, if it is possible for leakage to approach a dangerous level, the continued verification of isolation will be needed to assure personnel safety (see Figure 5.4). When service or maintenance is performed on a hydrogen system, for example, continuous air monitoring may be needed to ensure that hydrogen leakage does not result in a flammable atmosphere or oxygen deficiency.

Verification and Group Lockout/Tagout

Verification of energy isolation must also be assured when service or maintenance tasks involve multiple persons or groups ("group lockout/tagout"). In such situations, each authorized employee involved should generally affix his or her personal locks/tags to each isolating device and personally verify that the equipment has been de-energized and isolated.

With complex lockout/tagout operations that involve many employees and numerous energy isolating devices, however, this may not be practical or feasible. OSHA recognizes the need for flexibility during group lockout/tagout, and allows alternative group lockout/tagout procedures to be used as long as effective individual protection is maintained. For example, the employer may designate a primary authorized employee with overall responsibility for a group working under the protection of a group lockout/tagout device or devices (a "job lock" or "master tag"). The primary authorized employee may verify *for the group* that the machine or equipment has been effectively isolated from the hazardous energy sources. Each authorized employee participating in the group lockout/tagout, however, has the right to personally verify the effectiveness of the lockout measures if they so choose. Each authorized employee, or alternatively, the primary authorized employee, must verify isolation of hazardous energy sources to the extent necessary to ensure that the servicing/maintenance can be performed safely. Each applicable energy isolation device must be verified to assure effective energy isolation. Verification is also necessary by each group of workers before starting work at shift changes. See Chapter 6, *Group Lockout/Tagout* for more specifics on verification under group lockout/tagout.

Figure 5.4. Level controls with alarm.

Chapter 6

Group Lockout/Tagout

Simply stated, group lockout/tagout involves the performance of servicing or maintenance activities by more than one employee. Group lockout/tagout operations can range from simple activities involving a few persons over a single shift to complex operations involving crews, crafts, departments, and contractors spanning multiple shifts or even days. Various group lockout/tagout procedures and techniques can be used to provide effective protection for the group, including personal lockout/tagout of each isolating device, and the use of job locks, job tags, master tagging systems, work permits, and access control systems. The specific procedures selected may depend on the physical size and extent of the equipment being serviced/maintained, the number of employees involved, the number of energy isolating devices to be locked/tagged out, the accessibility of energy isolating devices, and personal preference.

In most situations, where the servicing or maintenance task can be completed in a single shift by a small number of persons, a procedure that involves the installation of individual locks and tags by each member of the group on all energy isolating devices provides the greatest assurance of protection for employees, and should be used. Under this basic approach, each servicing employee is responsible for the application and removal of his or her own lockout or

Figure 6.1. Hasp used to accept multiple locks.

Figure 6.2. Ball valve lockout accepts multiple locks.

tagout devices. Individual locks and tags can often be attached directly to the energy isolating device or the lockout device used to secure it. If necessary, a hasp or other device that accepts multiple locks can be attached to the energy isolating device (see Figures 6.1 and 6.2).

In complex situations, however, when the work involves many persons and/or extends across several workshifts, this approach may be impractical. OSHA recognized the need to provide employers with alternatives to having each employee lock and/or tag out each energy isolating device, and consequently, the language of the lockout/tagout standard permits the use of alternative procedures. When an alternative procedure is used, however, it must provide protection equivalent to that of individual lockout/tagout, and must otherwise comply with OSHA's provisions for energy control procedures.

1910.147 (f)(3)(i)

When servicing and/or maintenance is performed by a crew, craft, department or other group, they shall utilize a procedure which affords the employees a level of protection equivalent to that provided by the implementation of a personal lockout or tagout device.

For example, an authorized employee may secure an energy isolating device *for a group* with a job lock or similar procedure, as long as procedures are used to assure each individual's control of the energy hazards, such as using a lockbox for retention of the job lock key and to which each employee affixes his/her personal lock or tag. In a tagging system, a master tag may be used, as long as each employee personally signs on and signs off on it, and as long as the tag clearly identifies each authorized employee who is being protected by it.

Paragraph (f)(3) of the lockout/tagout standard also contains several key provisions for designating and assigning responsibility to authorized employees and for coordinating work forces that must be included in all group lockout or tagout procedures. Paragraph (f)(3)(ii)(A), for example, requires that primary responsibility for the safety of the group (or a set number of employees) be vested in an authorized employee. This employee is directly responsible for the performance of the overall servicing, and for the safety of all the employees in the group.

In addition, paragraph (f)(3)(ii)(B) requires the employer to develop and implement procedures for determining the exposure status of individual crew members and for taking appropriate measures to control or limit that exposure. OSHA sees these provisions as requiring at least the following steps:

1. Verifying shutdown and isolation of the equipment or process before allowing a crew member to place a personal lockout or tagout device on an energy isolating device or on a lockout box, board, or cabinet;

2. Ensuring that all employees in the crew have completed their assignments, removed their lockout and/or tagout devices from the energy isolating device, the box lid, or other device used, and are in the clear before turning the equipment or process over to the operating personnel or simply turning the machine or equipment on;

3. Providing the necessary coordinating procedures for ensuring the safe transfer of lockout or tagout control devices between other groups and work shifts.

> 1910.147 (f)(3)(ii)
> Group lockout or tagout devices shall be used in accordance with the procedures required by paragraph (c)(4) of this section including, but not necessarily limited to, the following specific requirements:
>
> (f)(3)(ii)(A)
> Primary responsibility is vested in an authorized employee for a set number of employees working under the protection of a group lockout or tagout device (such as an operations lock);
>
> (f)(3)(ii)(B)
> Provision for the authorized employee to ascertain the exposure status of individual group members with regard to the lockout or tagout of the machine or equipment;
>
> (f)(3)(ii)(C)
> When more than one crew, craft, department, etc. is involved, assignment of overall job-associated lockout or tagout control responsibility to an authorized employee designated to coordinate affected work forces and ensure continuity of protection.

Another key provision of the standard that must be incorporated into all group lockout or tagout procedures requires that each employee in the group affix a personal lockout or tagout system device as part of the group lockout.

> 1910.147 (f)(3)(ii)(D)
>
> Each authorized employee shall affix a personal lockout or tagout device to the group lockout device, group lockbox, or comparable mechanism when he or she begins work, and shall remove those devices when he or she stops working on the machine or equipment being serviced or maintained.

This provision was included in the lockout/tagout standard for several reasons:

- Placing a personal lockout or tagout device enables an employee to control his or her own protection, rather than having to depend upon another person.
- Using a personal lockout or tagout device enables each servicing employee to verify that the equipment has been properly de-energized in accordance with the energy control procedure, and to affix his or her device to indicate that verification.
- The presence of an employee's lockout or tagout device informs all other persons that the employee is working on the equipment.
- As long as that device remains attached, all employees know that the job is not completed and that it is not safe to reenergize the equipment.

- The servicing employee continues to be protected by the presence of the device until he or she removes it.

The use of personal locks/tags by involved employees also helps the authorized employee in charge of the group lockout to evaluate the current status of the servicing operation, and to determine which of the servicing employees are working on the equipment at a particular time.

OSHA Instruction STD 1-7.3 sets forth guidelines for the enforcement of the lockout/tagout standard by its inspectors when group lockout or tagging is involved. These guidelines can be summarized as follows:

1. Group lockout/tagout procedures must be tailored to the specific operation involved. Irrespective of the situation, the requirements of the final rule specify that each employee performing maintenance or servicing activities be in control of hazardous energy during his or her period of exposure.

2. The procedures must ensure that each authorized employee is protected from the unexpected release of hazardous energy by personal lockout or tagout devices. No employee may affix the personal lockout or tagout device of another employee.

3. The use of such devices as master locks and tags are permitted and can serve to simplify group lockout/tagout procedures. For example, a single lock may used on each energy isolating device, together with the use of a lockbox for retention of the keys and to which each authorized employee affixes his or her lock or tag (see Figure 6.3). In a tagging system, a master tag may be used, as long as each employee personally signs on and signs off on it and as long as the tag clearly identifies each authorized employee who is being protected by it.

4. All other provisions of the paragraph continue to apply.

In the following pages, several example procedures taken from Appendix C of OSHA Instruction STD 1-7.3 are presented to illustrate the implementation of a group lockout/tagout when multiple energy isolating devices and/or many servicing and maintenance personnel are

Figure 6.3. Group lock box.

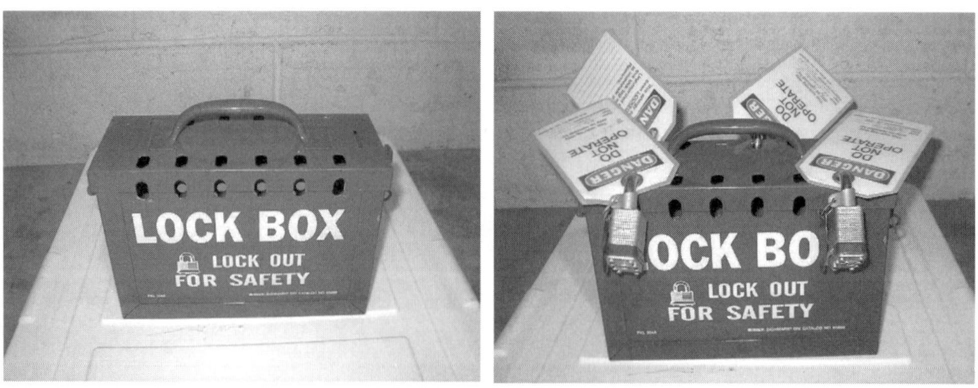

involved. A basic organizational structure for group lockout/tagout is also presented. First, however, several important terms that are unique to group lockout/tagout are defined.

Group Lockout/Tagout Definitions

Primary Authorized Employee. The authorized employee who has overall responsibility for adherence to the company lockout/tagout procedure. (See 29 CFR 1910.147(f)(3)(ii)(A).)

Principal Authorized Employee. An authorized employee who oversees or leads a servicing/maintenance group.

Job-Lock. A device used to ensure the continuity of energy isolation during a multi-shift operation. It is placed upon a lockbox. The assigned primary authorized employee from each shift controls the job-lock key.

Job-Tag with Tab. A special tag for tagout of energy isolating devices during group lockout/tagout procedures. The tag's tab is removed for insertion into the lockbox. The company procedure requires that the tagout job-tag cannot be removed until the tab is rejoined to it.

Master Lockbox. The lockbox into which all keys and tabs from the lockout or tagout devices securing the machine or equipment are inserted and which is secured by a job-lock during multi-shift operations.

Satellite Lockbox. A secondary lockbox to which each authorized employee affixes his or her personal lock or tag.

Master Tag. A document used as an administrative control and accountability device. This device is normally controlled by the operations department personnel and is a personal tagout device if each employee personally signs on and signs off on it and if the tag clearly identifies each authorized employee protected by it.

Work Permit. A control document that authorizes specific tasks and procedures to be accomplished.

Basic Group Lockout/Tagout Organizational Structure

The organization and coordination of the involved employees and groups is a critical success factor in the safe implementation of group lockout/tagout procedures. The following basic organizational structure described by OSHA provides an example approach:

a) A primary authorized employee is designated. This employee exercises primary responsibility for implementation and coordination of the lockout/tagout of hazardous energy sources for the equipment to be serviced.

b) The primary authorized employee coordinates with equipment operators before and after the completion of servicing and maintenance operations that require lockout/tagout.

c) A verification system is implemented to ensure the continued isolation and de-energization of hazardous energy sources during maintenance and servicing operations.

d) Each authorized employee is assured of his or her right to verify individually that the hazardous energy has been isolated and/or de-energized.

e) When more than one crew, craft, department, etc., is involved, each separate group of servicing/maintenance personnel is accounted for by a principal authorized employee from each group. Each principal employee is responsible to the primary authorized employee for maintaining accountability of each worker in that specific group in conformance with the company procedure. No person may sign on or sign off for another person, or attach or remove another person's lockout/tagout device, unless the provisions of the exception to 29 CFR 1910.147(e)(3) are met.

Example Group Lockout/Tagout Procedures

OSHA describes four basic types of procedures for implementing group lockout/tagout in their Instruction STD 1-7.3. These basic procedures can be used to address circumstances ranging from a small group of servicing/maintenance employees during a one-shift operation to a comprehensive operation involving many over a longer period.

- Type A. Each authorized employee places his or her personal lock or tag upon each energy isolating device and removes it upon departure from that assignment. Each authorized employee verifies or observes the de-energization of the equipment.
- Type B. Under a lockbox procedure, a job-lock or job-tag with tab is placed upon each energy isolation device after de-energization. The keys or removed tabs are then placed into a lockbox. Each authorized employee assigned to the job then affixes his or her personal lock or tag to the lockbox. As a member of a group, each assigned authorized employee verifies that all hazardous energy has been rendered safe. The lockout/tagout devices cannot be removed or the energy isolating device turned on until the appropriate key or tab is matched to its lock or tag.
- Type C. After each energy isolating device is locked/tagged out and the keys and tabs placed into a master lockbox, each servicing/maintenance group principal authorized employee places his or her personal lock or tag upon the master lockbox. Then each principal authorized employee inserts his or her key into a satellite lockbox to which each authorized employee in that specific group affixes his or her personal lock or tag. As a member of a group, each assigned authorized employee verifies that all hazardous energy has been rendered safe. Only after the servicing/maintenance functions of the specific subgroup have been concluded and the personal locks or tags of the respective employees have been removed from the satellite lockbox can the principal authorized employee remove his or her lock from the master lockbox.
- Type D. During operations to be conducted over more than one shift (or even many days or weeks) a system such as described here might be used. Single locks or tags are affixed upon a

lockbox by each authorized employee as described in Type B or Type C above. The master lockbox is first secured with a job-lock before subsequent locks by the principal authorized employees are put in place on the master lockbox. The job-lock may have multiple keys if they are in the sole possession of the various primary authorized employees (one on each shift). As a member of a group, each assigned authorized employee verifies that all hazardous energy has been rendered safe. In this manner, the security provisions of the energy control system are maintained across shift changes while permitting reenergization of the equipment at any appropriate time or shift.

In a group lockout/tagout operation, the primary authorized employee is also responsible for verifying that all elements of the group servicing have, in fact, been completed, and that it is safe to reenergize the system, before he or she removes the group device. Thus, the additional step provides further assurance that reenergizing the equipment will not endanger employees. Again, it is essential that the lockout/tagout devices not be removed by anyone else but the persons who applied them, except in emergencies.

Continuity of Lockout/Tagout Protection During Shift and Personnel Changes

Paragraph (f)(4) of the lockout/tagout standard requires that specific procedures be developed and used to ensure continuous protection of employees and an orderly transfer of lockout/tagout protection during shift or personnel changes.

> 1910.147 (f)(4)
>
> *Shift or personnel changes.* Specific procedures shall be utilized during shift or personnel changes to ensure the continuity of lockout or tagout protection, including provision for the orderly transfer of lockout or tagout device protection between off-going and oncoming employees, to minimize exposure to hazards from the unexpected energization or start-up of the machine or equipment, or the release of stored energy.

To accomplish this uninterrupted protection for involved employees, it is essential that the applied locks/tags be maintained on energy isolating devices throughout the transition period. As with group lockout or tagout, the specific method of accomplishing this task must be part of the company's written energy control program.

There are a number of ways to achieve the orderly transfer and continuity of protection during shift or personnel changes. A basic method involves the application of locks/tags by incoming shift employees before outgoing shift employees remove their devices. Under this approach, often referred to as the "handshake" approach, employees of both shifts meet first to discuss the specifics and status of the service or maintenance task, and the incoming shift employees assure that the equipment is in a safe condition to be worked on by verifying that all energy sources are effectively controlled and isolated. In this manner, the incoming shift employees accept control of the service or maintenance task prior to the release of control by the outgoing employees. Consequently, there is no gap in coverage between the outgoing employee's removal of his or her lockout or tagout device and the incoming employee's attachment of his or her own device.

Another system involves the application of an operations or job lock. An operations lock, for example, can be applied to all energy isolating devices by a primary authorized employee; they are the first locks attached to the equipment when the equipment is de-energized, and are the last locks removed when the job is completed. Each servicing employee then attaches his or her personal locks/tags to the isolating devices (using a hasp if necessary) while working on the equipment, and removes the device when their job is completed, or when leaving for the day. The primary authorized employee's locks/tags on the isolating devices ensure that the equipment remains isolated for the duration of the service or maintenance task. As with the handshake approach, employees inspect the equipment before proceeding with the service or maintenance task, verify that the equipment has been de-energized and isolated, and ensure that proper procedures have been followed.

In a similar approach, the primary authorized employee applies locks/tags to all energy isolating devices, and places the keys to all the operations locks in a lockbox. The primary authorized employee then places an operations lock on the lockbox, which is the first lock attached to the lockbox, and is the last lock removed when the job is completed. Each servicing employee then attaches his or her personal locks/tags to the lockbox while working on the equipment, and removes them when their job is completed, or when leaving for the day. Again, employees inspect the equipment and verify that the equipment has been effectively de-energized and isolated before proceeding with the service or maintenance task. With this approach, each authorized employee is afforded a means to use his or her personal lockout or tagout devices so that no single employee has control of the means to remove the group lockout or tagout devices while employees are still servicing the machine or equipment. The primary authorized employee's locks/tags on the isolating devices ensure that the equipment remains isolated for the duration of the service or maintenance task. The locks/tags of the other authorized employees on the lockbox ensure that the keys to the isolating devices cannot be accessed until all personal devices are removed, assuring individual protection. For more complex tasks, a variation of this approach, such as that described under Type D above, can also be used to ensure continued protection during shift or personnel changes. Often, a log is kept to account for personnel and to track the application and removal of locks and tags and work progress.

Contractors and Outside Servicing Personnel

Paragraph (f)(2) of the lockout/tagout standard addresses coordination of lockout/tagout procedures between company and contractor employees. The provisions of this section require that whenever outside servicing personnel (contractor employees) perform covered service or maintenance activities on the company's equipment, each employer must inform the other employer of their respective lockout or tagout procedures. Each employer must also ensure that his or her own employees understand and comply with the restrictions and prohibitions of the energy control program in use.

1910.147 (f)(2)

(f)(2)(i)

Whenever outside servicing personnel are to be engaged in activities covered by the scope and application of this standard, the on-site employer and the outside employer shall inform each other of their respective lockout or tagout procedures.

(f)(2)(ii)

The on-site employer shall ensure that his/her employees understand and comply with the restrictions and prohibitions of the outside employer's energy control program.

Close coordination of procedures between company and contract employees is important because of the potential for misunderstandings that could lead to the injury of either or both company's employees. For example, the contractor's employees might be familiar with specific equipment being serviced, but they might not be familiar with the energy control procedures in use at the given facility. Similarly, the company's employees at the facility might be familiar with the procedures in use at the facility, but not the contractor's, which might differ from their own. If the energy control procedures are not coordinated, each group of employees might be endangered by the actions of the other, even if each one followed its own procedures. For example, not following the same sequence of procedures could potentially result in employees of one group being exposed to hazardous energy while the other group is attempting to verify de-energization with a deliberate start-up attempt. Other problems could arise if one group uses lockout while the other group uses tagout alone.

The provisions of section (f)(2) are intended to prevent misunderstandings by either the company or contractor employees by requiring that each employer inform the other employer of the procedures used by his or her employees. In this regard, each set of employees should acknowledge that they understand and will comply with the restrictions and prohibitions of the energy control program in use. For example, if there are elements of the contractor's procedures that need to be explained to the facility employees, or if there are other steps needed to assure the safety of the contractor's employees, the facility employer must provide his or her employees with the information to ensure the necessary protection. Ideally, both groups of employees should meet and agree upon the specific procedures to be followed.

Various aspects of a contractor's energy control program should be examined to assure that his or her employees are not placed at an increased risk. Questions to ask include:

- Is the contractor's means of notifying the affected employees of the pending lockout or tagout as thorough as the facility employer's?
- Is the procedure for identifying the energy isolating devices as exhaustive or complete as the facility employer's?

- Is the method of lockout or tagout used by the contractor recognized and respected by the facility's employees?
- Does the contractor's procedure take into account the possibility of reaccumulation of stored energy (if that is a potential problem)?
- Does the contractor's procedure for removal of lockout or tagout devices and reenergizing and startup of the machine or equipment provide for employee notification and ensuring the equipment is safe before startup?

If any of the steps in the contractor's procedures fail to cover significant or essential conditions of the workplace which could adversely affect the safety of the facility employees, action must be taken by the facility employer to minimize the potential for injury to his or her employees. Regardless of the degree of coordination, each employer has an independent obligation to provide adequate protection for his or her own employees.

Group Tagout

When the service or maintenance task involves equipment that is not capable of being locked out, OSHA advises that it is possible to extend coverage of group servicing activities under paragraph (f)(3) to include tagout by requiring each servicing employee to attach his or her own device, as long as the tagout system provides full employee protection as set forth in paragraph (c)(3) of the lockout/tagout standard. Additional training is also required for employees using tagout systems alone, in accordance with section (c)(7)(ii) of the standard.

In a master tagging system, a master tag is attached to the isolating devices. The master tag must clearly identify each authorized employee who is being protected by it, and each employee must personally sign on and sign off on it. The master tagging system must also incorporate additional measures as necessary to provide a level of safety equivalent to that obtained by using a group lockout program.

In situations during shift changes where the equipment is to remain de-energized throughout the transition, and the outgoing employee must remove his or her lock/tag before the incoming employee arrives, it is also possible to establish a procedure that involves use of an interim tag by which an employee transfers servicing duties to an employee on the next shift. In this situation, the procedure could allow for the outgoing employee to apply an interim tag at the time he or she removes his or her device, indicating that the lock/tag had been removed, but that the machine or equipment had not been reenergized. The incoming employee verifies that the system is still de-energized, and removes the interim tag and substitutes his or her own personal lock/tag. This ensures that the continuous protection is maintained from one shift to another.

When tagout devices alone are used, it is possible to use a tag with space for the outgoing employee to sign off, giving the date and time, and for the incoming employee to sign on, also giving the date and time. Each employee verifies the de-energization and energy isolation for his or her own protection before signing onto the tag. Again, the procedure must provide full employee protection and must be documented in the employer's energy control program.

Group Lockout/Tagout Procedures for Complex Operations

The basic types of group lockout/tagout procedures described previously require the placement of individual lockout/tagout devices by each authorized employee on individual energy isolating devices or on a group lockout device, such as a lockbox. Servicing and maintenance of sophisticated and complex equipment, however, may involve large numbers of energy isolating devices, employees, and multiple shifts (such as found in petroleum and chemical production). Modification of these basic procedures or implementation of a more feasible system may be needed. Energy control procedures in these situations are often supplemented by the use of work permits, access control programs, or other systems of continuous worker accountability.

In these complex situations, OSHA acknowledges that the removal and replacement of the lockout/tagout devices each shift could be overly burdensome. When the complexity of the servicing operation necessitates an alternative to such frequent attachment and removal of lockout or tagout devices, OSHA advises that the use of the work permit or comparable means, with each employee signing in and out as he or she begins or stops working on the equipment, combined with the servicing employees verifying that the equipment is de-energized prior to beginning work, is an acceptable approach to compliance with group lockout/tagout and shift change provisions of the lockout/tagout standard. Generally, however, such systems should not be used unless it can be demonstrated that conventional group lockout procedures are not feasible.

OSHA's Instruction STD 1-7.3 provides further insight on how a group lockout/tagout procedure can be implemented in complex situations:

a) Complex process equipment that is scheduled for servicing/maintenance operations is generally identified by plant supervisors who then issue specific work orders regarding the operations to be performed.

b) In most instances where complex process equipment is to be serviced or maintained, the process equipment operators can be expected to conduct the shutdown procedure. This is generally due to their in-depth knowledge of the equipment and the need to conduct the shutdown procedure in a safe, efficient, and specific sequence.

c) The operations personnel normally prepare the equipment for lockout/tagout as they proceed and identify the locations for blanks, blocks, etc., by placing operations locks and/or tags on the equipment. The operations personnel can be expected to isolate the hazardous energy, and drain and flush fluids from the process equipment following a standard procedure or a specific work permit procedure.

d) Upon completion of shutdown, the operations personnel review the intended job with the servicing and maintenance crews and ensure their full comprehension of the energy controls necessary to conduct the servicing or maintenance safely. During or immediately after the review of the job, the servicing and maintenance crews install locks, tags and/or special isolating devices at previously identified equipment locations following the specified work permit procedure.

e) Line openings necessary for the isolation of the equipment are normally permitted only by special work permits issued by operations personnel. (Such line openings should be monitored by operations personnel as an added safety measure.)

f) All of the previous steps should have been documented by a master system of accountability and retained at the primary equipment control station for the duration of the job. The master system of accountability may manifest itself as a master tag, which is subsequently signed by all of the maintenance/servicing workers if they fully comprehend the details of the job and the energy isolation devices actuated or put in place. This signing by the respective workers further verifies that energy isolation training relative to this operation has been conducted.

g) After the system has been rendered safe, the authorized employees verify energy controls (see discussion under *Verification*).

h) Specific work functions are controlled by work permits that are issued for each shift. Each day, each authorized employee assigned must sign in on the work permit at the time of arrival to the job and sign out at departure. Signature, date, and time for sign-in and sign-out are recorded and retained by the applicable crew supervisor who, upon completion of the permit requirements, returns the permit to the operations supervisor. Work permits could extend beyond a single shift and may subsequently be the responsibility of several supervisors.

i) Upon completion of the tasks required by the work permit, the authorized employees' names can be signed off the master tag by their supervisor once all employees have signed off the work permit. The work permit is then attached to the master tag. (Accountability of exposed workers is maintained.)

j) As the work is completed by the various crews, the work permits and the accountability of personnel are reconciled jointly by the primary authorized employee and the operations supervisor.

k) During the progress of the work, inspection audits are conducted.

l) Upon completion of all work, the equipment is returned to the operations personnel after the maintenance and servicing crews have removed their locks, tags, and/or special isolating devices following the company procedure.

m) At this time all authorized employees who were assigned to the tasks are again accounted for and verified to be clear from the equipment area.

n) After the completion of the servicing/maintenance work, operations personnel remove the tags originally placed to identify energy isolation.

o) Operations personnel then begin check-out, verification, and testing of the equipment prior to its being returned to production service.

Verification and Group Lockout/Tagout

Verifying equipment de-energization and isolation is arguably the most important element for assuring continuity of protection during group lockout procedures. In general, each authorized employee involved in the group procedure should personally verify that the equipment has been effectively de-energized and isolated whenever feasible. The underlying rationale for this is that the incoming employee should not have to depend on the actions of another employee (especially one who has left the workplace for the day) for assurance that it is safe to work on the machine or equipment.

With complex lockout/tagout operations where this is not feasible and there is a need for flexibility, however, OSHA advises that alternative group lockout/tagout procedures can be used as long as effective individual protection is maintained. For example, where a group is working under the protection of a group lockout/tagout device such as a job lock or master tag, and the employer has designated a primary authorized employee with overall responsibility for the group, the primary authorized employee may verify *for the group* that the machine or equipment has been effectively isolated from the hazardous energy sources. This must be accomplished *before* authorized employees participating in the group lockout/tagout affix their personal lockout or tagout device to the group lockout/tagout box and *before* they perform servicing/maintenance activities. Each authorized employee participating in the group lockout/tagout, however, must be assured of their right to *personally* verify the effectiveness of the lockout measures, if they so choose. An authorized employee who chooses to verify the effectiveness of the isolation measures must perform this verification *after* affixing his or her personal lockout or tagout device to the lockbox and *before* performing servicing/maintenance activities.

Each authorized employee, or alternatively, the primary authorized employee, must verify isolation of hazardous energy sources to the extent necessary to ensure that the servicing/maintenance can be performed safely. Each applicable energy isolation device must be verified to assure effective energy isolation. This may involve a walk-through of the affected work area with operations personnel to verify effective isolation prior to beginning work. It is not sufficient for authorized employees and primary authorized employees to merely review tests in a job briefing and to rely on a locked lockbox. Where multiple groups are involved, verification is also necessary by each group of workers before starting work at shift changes.

Chapter 7

Vehicles and Equipment

Workers who perform service, maintenance, and repairs on automobiles, buses, trucks, and other motorized equipment encounter a variety of hazardous energy types, and face a range of potential hazards that could cause injury or death. Tasks that require workers to crawl or work under vehicles and equipment, for example, may expose them to potential energy (due to gravity) that poses an injury risk if the vehicle were to unexpectedly fall or roll. Other energy-related hazards include contact with exposed moving parts such as fans, belts, pulleys, and driveshafts, burns from contact with hot coolant and other fluids, and electrical shock or burns due to contact with electrical components.

The following case study provides an example of the potential hazards associated with work on vehicles and equipment. A truck driver was crushed and killed between the frame and dump box of a dump truck. Apparently a safety overtravel cable attached between the truck frame and the dump box malfunctioned by catching on a protruding nut of an air brake cylinder. This prevented the dump box from being fully raised, halting its progress at a point where about 20 inches of space remained between it and the truck frame. The employee, apparently assuming that releasing the cable would allow the dump box to continue upward, reached between the rear dual wheels and over the frame, and disengaged the cable with his right hand. The dump box then dropped suddenly, crushing his head. The employee had not received training or instruction in proper operating procedures and was not made aware of all potential hazards in his work. Had blocking or bracing been used by the employee to secure the dump body, this fatality might have been prevented (see Figure 7.1).

Figure 7.1. OEM blocking device for use when working under raised dump body.

Application of the Lockout/Tagout Standard to Motor Vehicles and Equipment

While there is no specific reference to the servicing and maintenance of motor vehicles in the lockout/tagout standard, OSHA advises that the standard applies to all general industry workplaces, and that where there is a potential for employee injury as a result of unexpected energization, start-up, or release of stored energy during the servicing and maintenance of motor vehicles, the provisions of the lockout/tagout standard apply.

With respect to construction equipment such as front-end loaders, bulldozers, backhoes, graders, dump trucks, augers, pile drivers, and mobile cranes, servicing and maintenance tasks that involve the positioning of components such as buckets, blades, and machine body parts, can also present serious hazards for personnel, and safe work practices (such as blocking of raised parts) are clearly needed to prevent injury. Coverage of the lockout/tagout standard for such activities, however, depends on who is performing the service or maintenance task.

For example, since OSHA exempted the construction industry from the scope of the final standard, lockout/tagout requirements do not apply to such activities if they are done by a *construction* company. If these tasks were carried out by a general industry workplace such as an automotive service or repair shop, however, the lockout/tagout standard's provisions would apply. Regardless of the exemption, all employers that use such equipment should implement whatever measures are necessary to protect employees engaged in such activities.

Insight on OSHA's position relative to the type of motor vehicles covered by the lockout/tagout standard can be found in the discussion on page 36657 of the preamble to the final rule. The pertinent portion of the discussion reads as follows:

"Of additional concern in the imposition of regulations in the construction industry is the uniqueness of earthmoving equipment, such as lattice boom mobile cranes, front-end loaders, bulldozers, scrappers, and dump trucks. As opposed to maintenance on automobiles, buses and over-the-road trucks where the removal of the ignition key usually ensures that the engine cannot be restarted and the vehicle may be worked on, some of the maintenance of the above mentioned earthmoving equipment involves the positioning of components, such as buckets, blades and machine body parts which present extraordinary hazards and the means to minimize the potential for injury to employees involve additional considerations which were not adequately addressed during the course of the rulemaking proceeding."

For general industry workplaces, therefore, if service or maintenance tasks on motor vehicles and equipment pose potential injury risks to employees due to unexpected start up or energization, then the safe performance of those tasks should be addressed in the employer's energy control program.

Energy Control for Motor Vehicles and Equipment

To prevent injury due to contact with exposed moving parts such as fans, belts, pulleys, and driveshafts, the vehicle or equipment should be shut down or turned off when possible during the service or maintenance, and all moving parts should be allowed to come to a complete stop. As with industrial machinery, the accidental restart of vehicles and equipment and movement of auxiliary parts or attachments is a potential concern to be addressed. Serious injury could occur if the vehicle or equipment's motor or engine were to be accidentally started up while a servicing employee had a hand or other body part in a danger zone. Injury could occur, for example, if a worker's hand were struck by fan blades or pinched between a belt and pulley.

If employees perform service or maintenance tasks that put them at risk of injury if a vehicle were accidentally started, measures must be implemented pursuant to the lockout/tagout standard to provide effective employee protection. In this regard, OSHA advises that removal of the ignition key generally eliminates most of the potential for injury as a result of unexpected energization, and is an acceptable form of lockout as long as this action effectively protects employees. If the key is removed, it must be kept under the exclusive control of the employee performing the work. Procedures for controlling access to keys (for example, using a key cabinet), along with employee instruction on those procedures should also be implemented.

OSHA also advises that removal of the ignition key alone will not adequately protect employees from accidental start up in all cases, and other measures such as disconnection of the positive battery cable may be needed. For example, there are several types of equipment on vehicles that can operate independently of the ignition system, most notable of which are engine cooling fans and air bags. Engine cooling fans may be mechanical (belt-driven) or electrical. Even after the ignition is turned off, electrical engine fans may continue to run, or may start up automatically. Similarly, a vehicle's airbag could potentially injure a person working inside the vehicle if accidentally deployed. If the service or maintenance task exposes the worker to accidental deployment of the airbag, the manufacturer's instructions for disabling the system should be followed.

Figure 7.2. Pyramid jack used to support raised vehicle.

To prevent movement of the vehicle, the brakes should be set and wheels should be chocked as necessary. If the vehicle is raised to perform work underneath, jacks, stands, or other blocking devices must be used to prevent the vehicle from falling in case the hoist or jack fails or the control levers or pedals are inadvertently struck and the load released (see Figure 7.2). When a person is working under a blocked-up vehicle, other employees should not engage in any activities that may cause the car to fall off its jacks.

When an automotive lift is used to raise a vehicle to perform work, the lift should be equipped with appropriate safety devices such as blocking pins or ratchet mechanisms to prevent the lift and/or suspended vehicle from falling due to hydraulic or mechanical failure. Workers should verify that the safety device is properly engaged and working before working beneath the raised vehicle. If necessary, drivetrain components such as transmissions, transfer cases, and axles should be supported to prevent them from falling.

To protect against burns, hot parts such as the engine block, radiator, and exhaust components should be allowed to cool if possible, or insulating gloves or blankets should be used for protection. To check or fill a radiator, heavy work gloves should be used to remove the cap.Typically, the cap is first moved to a safety catch found at a half-turn, where any steam can be bled off, before removing the cap. Care should also be exercised when using gasoline or solvents near hot engines or electrical components; a spill can result in a serious fire.

Diagnostic Procedures and Adjustments

Making adjustments and repairs to vehicles and equipment while the engine is running can be dangerous, particularly when work is done near fans, clutches, and belts. The lockout/tagout standard requires that equipment and machinery be shut down or turned off while performing service or maintenance. Consequently, work done near exposed moving parts should done with the engine stopped when possible. An alarming number of injuries occur each year to people attempting to clear jams from equipment such as snowblowers and wood chippers (see Figure 7.3) by hand. Common sense dictates that the equipment be shut down before attempting such a task, and that a tool rather than one's hand be used for such a purpose.

Adjustments, OSHA advises, must be made while the equipment is de-energized, unless absolutely necessary. If this is not possible, however, such as is the case with certain diagnostic or other procedures that require the engine to be running, alternative steps must be taken to protect employees from harm. Such diagnostic procedures and adjustments should also follow the sequence of actions outlined in paragraph (f)(1) of the lockout/tagout standard for testing

Figure 7.3. Wood chipper.

or positioning of machines, equipment, or components. An example of an adjustment to a motor vehicle that must be made with the engine running is adjustment of the timing.

In those cases where it is not possible to de-energize the vehicle or equipment, OSHA advises that to protect the employees the machine guarding and safe work practice requirements of Subparts O and M of the General Industry Standards must be implemented to the extent feasible. Guards (such as the fan shroud) may only be removed if it is necessary to do so in order to perform the diagnostic procedures, and must be replaced as soon as possible. Employees should keep hands and other body parts away from moving parts to the extent possible. Workers should not wear jewelry or loose-fitting clothing that could be caught in moving parts, especially loose-fitting sleeves or cuffs, neckties, and so on. Personal protective equipment such as eye protection must also be used as needed.

Stored Energy in Vehicles and Equipment

In addition to potential energy associated with raised or suspended parts, other forms of potentially hazardous stored or residual energy may also be encountered on various vehicles, equipment, attachments, and accessories, including stored electrical energy in batteries and capacitors, pressure in hydraulic and pneumatic systems, and compressed or extended springs. When necessary, this energy must be relieved, dissipated, disconnected, restrained, or otherwise rendered safe to allow the service or maintenance task to be performed safely. This can be done, for example, by blocking components that could move or fall, by bleeding down hydraulic and/or pneumatic systems, and by moving components to a point at which springs are at a neutral (relaxed) state *(see Chapter 3 for more on Methods of Energy Control)*. Again, shutting down and turning off the vehicle or equipment will generally be the first step in preparation for service or maintenance.

To control electrical energy, removal, tagging and/or locking of battery leads may also be necessary, and can often be accomplished with a plug lockout device. Care should be taken to prevent arcing while batteries are being charged, tested, or handled. With hydraulic systems, if the residual hydraulic energy in auxiliary parts or attachments is not properly relieved, dissipated, or otherwise restrained, an accident could occur if controls were to be inadvertently activated (for example, by another employee not realizing that the servicing employee was exposed). If the work is to be done on or near moving parts and there is a chance that the controls could be inadvertently activated, the controls should be locked/tagged out.

Various types of springs are associated with vehicle suspensions, including coil springs, leaf springs, and torsion springs. In some cases, moving the axle to a position in which the spring is in a relaxed state may be sufficient to dissipate the potential energy in the spring. In other cases, special tools may be needed to compress, relax, or secure the spring.

Work on construction equipment such as lattice boom mobile cranes can also be hazardous. Many workers have been killed when the boom fell and crushed them after they removed the main pins holding the boom to the base of the crane. The cause of these fatalities was the failure of workers to properly block and secure the boom to prevent it from falling during disassembly (see Figures 7.4 and 7.5).

Figure 7.4 (above). Boom lowered on lattice boom mobile crane in preparation for service.

Figure 7.5 (right). Main pin connecting lattice boom to crane base.

The following are some general guidelines for the safe performance of service and maintenance on construction equipment taken from the National Safety Council's *Accident Prevention Manual for Industrial Operations*:

- On construction equipment, repairs should be made with the component in a safe position.
- When repairs are to be made, set the brakes, secure the boom, lower the bucket to the ground, and take the machine out of gear to prevent accidental movement.
- Before repairing the vehicle, notify the operator about the nature and location of the repair.
- Parts that must be in motion while workers are working on them should be turned slowly, by hand if possible, in response to guidance or signals if multiple persons are involved. This precaution applies particularly to persons working around gears, sheaves, and drums. Workers who grasp ropes just ahead of the sheaves risk having their hands jerked into the sheaves. To prevent hand injuries, a rope being wound on a drum should be guided with a bar.
- If guards must be removed for convenience in making repairs, the job cannot be considered complete until the guards, plates, and other safety devices have been replaced.
- When cutting edges are to be replaced, the scraper bowl or dozer blade should always be blocked up. After the scraper has been lifted to the desired height, blocks are placed under the bottom near the ground plates. Apron arms are raised to the extreme height, and a block is placed under each arm, so the apron can drop enough to wedge each block firmly in place.

Before receiving wire rope on a drum or through sheaves, the operator should disengage the master clutch, idle the engine, and lock the brakes. The engine should be at a complete stop before working with the rope on a front-mounted drum.

Chapter 8

Energy Control Procedures

Under OSHA's lockout/tagout standard, employers are required to establish an energy control program to protect employees from energy-related hazards during service or maintenance. To provide guidance to employees for putting the program into practice, a written policy should be developed. The policy should define the purpose, scope, and application of the program, prescribe the responsibilities of managers, supervisors, and employees for implementing program elements, and describe the *general* policies and procedures for implementing lockout/tagout.

In addition to the written policy's general rules and procedures, more detailed equipment-specific energy control procedures (ECPs) also need to be prepared. The ECPs should work in conjunction with the overall policy, and supplement its general provisions with more specific guidance for locking/tagging out particular machines, equipment, or processes.

To begin the process of developing ECPs, a survey of the workplace needs to be conducted to identify the equipment that uses hazardous energy; ECPs are needed for all machines, equipment, and processes that may expose employees to energy-related hazards during service or maintenance. An individual (or individuals) then needs to be assigned to write the ECPs. Often, this task falls to supervisors with guidance from the company safety representative. In larger companies with numerous or complex machines and equipment, work teams or task groups may need to be assigned this task. Consultants can also help companies develop the needed ECPs if the necessary in-house knowledge or resources is lacking.

The personnel assigned to write and/or approve the ECPs should be qualified to perform these tasks. For example, these individuals should be familiar with the construction and operation of the equipment, and versed in the requirements of the energy control program. Persons developing ECPs must have the necessary knowledge to prescribe the proper steps for the safe application, use, and removal of lockout/tagout controls in the ECP. It is also important for them to have a good understanding of the service and maintenance tasks that are performed on the equipment. To achieve employee buy-in to use the procedures, the input of the employees who operate the equipment and the authorized employees who service it should be sought. Although not required by the standard, such active employee involvement enhances understanding of the procedures, and fosters a sense of ownership.

When Are ECPs Required?

Written ECPs are generally required whenever employees perform service or maintenance tasks that potentially expose them to injury due to unexpected energization or start-up of the equipment, or the release of stored energy. According to interpretive guidance provided by OSHA, the ECPs may be developed on an as-needed basis, but must be in place and followed *prior* to actually beginning the service or maintenance work, even if that work is of an emergency nature. Anticipating the need for and preparing ECPs for every potential service or maintenance task that may arise may be impractical, and for infrequent or non-routine situations, preparation of ECPs on an as needed basis may be appropriate. For machines, equipment, and processes that undergo routine service or maintenance, or where the need to implement energy control procedures on an emergency basis can be reasonably anticipated, the most prudent approach is to simply prepare the ECPs in advance.

The requirements for ECPs are set forth in section (c)(4) of the standard:

> 1910.147 (c)(4) Energy control procedure.
>
> (c)(4)(i)
>
> Procedures shall be developed, documented and utilized for the control of potentially hazardous energy when employees are engaged in the activities covered by this section.

Typically, ECPs need to be developed for the servicing or maintenance of most industrial machines and equipment. In drafting the lockout/tagout standard, however, OSHA recognized certain limited situations in which the development of specific, written ECPs would not contribute markedly to the overall safety of employees, and provided an exception from documenting the ECP when specific conditions can be met. These situations include work on machines and equipment that involves a single source of hazardous energy that is easily identified and isolated, with no potential for stored or residual energy. Since only a single energy source needs to be isolated, the procedure for controlling that energy is greatly simplified. In such cases, OSHA believes that the authorized employee can perform the procedure without referring to a written document.

The exception is found in a note to section (c)(4)(i):

> 1910.147 (c)(4)(i)
>
> Note: Exception: The employer need not document the required procedure for a particular machine or equipment, when all of the following elements exist:
>
> (1) The machine or equipment has no potential for stored or residual energy or reaccumulation of stored energy after shut down which could endanger employees;

(2) the machine or equipment has a single energy source which can be readily identified and isolated;

(3) the isolation and locking out of that energy source will completely de-energize and deactivate the machine or equipment;

(4) the machine or equipment is isolated from that energy source and locked out during servicing or maintenance;

(5) a single lockout device will achieve a locked-out condition;

(6) the lockout device is under the exclusive control of the authorized employee performing the servicing or maintenance;

(7) the servicing or maintenance does not create hazards for other employees; and

(8) the employer, in utilizing this exception, has had no accidents involving the unexpected activation or re-energization of the machine or equipment during servicing or maintenance.

The exception is intended to apply to situations in which the service or maintenance task can be carried out under lockout/tagout without detailed instructions. The exception does *not* exempt the equipment from being locked/tagged out; it only exempts the employer from having to prepare a written ECP for that equipment. The following example, provided by OSHA in the standard's preamble, describes an example situation where an ECP would not be required. "A motor in a small machine shop is wired into a single electrical disconnect with no other energy source, and the motor does not present the hazards of stored or residual energy. When the motor needs repair, the authorized employee can isolate the motor from the single energy source and lock it out, using his/her personal lockout device on the disconnect, in accordance with the procedures set forth in the standard. Under these conditions, and provided that no other employees are exposed to hazards from the servicing operation, the servicing may be performed without the need to document the energy control procedure."

All eight of the specified conditions must be met for the machine or equipment to qualify for the exception. While there is a lot of single energy source equipment that is eligible for the exception, it may be difficult or impossible for many industrial machines to meet all eight of the conditions. In these situations, the most prudent approach is to simply prepare ECPs for the equipment. In addition, if an employer, in claiming the exception, has an accident involving the machine or equipment in which hazardous energy is a factor, a documented ECP then becomes necessary.

What's Required in an ECP?

The OSHA standard also requires that the ECPs, where required, clearly and specifically outline the steps to be followed, techniques to be used, and measures to be applied by the employer for energy control. The procedure must specify the information that the authorized employees must know to accomplish the lockout/tagout on the particular machine, equipment, or process.

This includes specific steps to shut down, isolate, block, secure, place, and remove lockout or tagout devices, and to determine the effectiveness of those devices.

The required content for ECPs is prescribed in section (c)(4)(ii) of the standard:

> (c)(4)(ii)
>
> The procedures shall clearly and specifically outline the scope, purpose, authorization, rules, and techniques to be utilized for the control of hazardous energy, and the means to enforce compliance including, but not limited to, the following:
>
> (c)(4)(ii)(A)
>
> A specific statement of the intended use of the procedure;
>
> (c)(4)(ii)(B)
>
> Specific procedural steps for shutting down, isolating, blocking and securing machines or equipment to control hazardous energy;
>
> (c)(4)(ii)(C)
>
> Specific procedural steps for the placement, removal and transfer of lockout devices or tagout devices and the responsibility for them; and
>
> (c)(4)(ii)(D)
>
> Specific requirements for testing a machine or equipment to determine and verify the effectiveness of lockout devices, tagout devices, and other energy control measures.

It is common practice, in meeting the requirements of this section, to use the written policy to describe the general elements (the scope, purpose, authorization, rules, techniques and so on), and the ECPs to address the procedural steps specific to the particular equipment to be worked on [the requirements of (c)(4)(ii)(A) –(D)]. This approach can help keep the ECPs from becoming overly complicated. When appropriate, however, a note should be provided in the ECP indicating that it is to be used in conjunction with the provisions of the written lockout/tagout policy.

The detail required in the ECP depends on the complexity of the equipment involved, the type of service or maintenance to be performed, and the control measures to be used. For example, minimal detail is required in an ECP for a straightforward service or maintenance task that involves simply shutting down the machine at the controls, locking out an electrical disconnect, and attempting a restart to verify de-energization. In contrast, an ECP for a service or maintenance task that entails multiple energy sources, multiple machine components, complex isolation, and/or special verification steps, requires considerable detail. Often, complex ECPs can be simplified by referencing procedural steps that are adequately documented elsewhere. For example, operations and maintenance manuals or other system operating procedures may already detail special procedures for equipment shutdown or start-up that can be incorporated by reference.

In addition to the minimum requirements set forth in section (c)(4)(ii) (A)–(D), ECPs should also address the:

- machine, equipment, or process to be locked/tagged out;
- service/maintenance tasks to be performed;
- types and magnitude of hazardous energy associated with the machine, equipment, or process;
- type and location of the isolating devices;
- personnel involved;
- specific steps for re-energizing the machine and restoring it back to normal operation.

If not addressed in the written policy, or to provide added emphasis, the *general* steps for safely restoring equipment back to operation after the service or maintenance task is completed should be included in the ECP. These steps are outlined in paragraph (e)(1) and (2) of the standard, and require that the work area be inspected when the service or maintenance task is complete to ensure that nonessential items have been removed, and that the equipment is operationally intact and in a condition to be safely re-energized. The work area inspection must also ensure that all employees are safely positioned, and affected employees must be notified that lockout/tagout devices have been removed and that the equipment is to be re-energized. The example ECPs at the end of this chapter provide example language for addressing these required actions.

Although the OSHA standard does not expressly require the written ECP to address the *specific* steps or sequence for restoring the equipment back to service following completion of the service or maintenance task (beyond the removal of locks/tags), it is a good idea to do so when appropriate. With certain equipment, for example, it may be critical to follow a specified sequence for re-energization to avoid potential equipment damage. Electrical breakers and disconnects, for instance, should not be closed under load, or used to bypass equipment controls to directly start up equipment. The need for such procedures is addressed in NFPA 70E – *Standard for Electrical Safety Requirements for Employee Workplaces*:

> NFPA 70E, Ch. 4 (B)(6)
>
> Procedures for Restoring Electric Service. Procedures for restoring electric circuits and equipment to service shall be included in the lockout-tagout procedure...

An example ECP format, along with directions for completing it, are provided at the end of this chapter. Example ECPs for several types of equipment are also provided.

Generic ECPs

In section (c)(4)(ii) of the standard, OSHA repeatedly uses the word "specific" when detailing the required elements of the energy control procedure. This was done to emphasize the need to have a detailed procedure that clearly and specifically outlines the steps to be followed by authorized employees when locking/tagging out machines, equipment, and processes. In general, the provisions of section (c)(4)(ii) should be interpreted to mean that a specific procedure should be prepared for every *different* machine or piece of equipment covered by the standard. According to interpretative guidance provided in OSHA's Instruction STD 1-7.3, however, *similar* machines and/or equipment can be covered with a single procedure if it adequately addresses the hazards to be encountered and steps to be taken during service or maintenance for all the machines and/or equipment covered by the procedure. Similar machines and equipment, for example, may include those using the same type and magnitude energy and the same or similar types of energy controls. Generic procedures, therefore, can be applied to similar equipment where the hazards, shut down procedures, energy controls, and so on, are identical for each piece of equipment; where this is not the case, however, individual procedures are required.

Further insight on OSHA's view of this matter is provided in a September 19, 1995 letter of interpretation:

> "[OSHA agrees] that a separate procedure does not have to be developed for each and every machine or piece of equipment and that a comprehensive (generic) energy control procedure with supplemental checklists or appendices may very well address adequately the steps necessary to perform servicing and maintenance safely. Such a procedure would normally be considered one procedure. OSHA believes, however, that there are also situations which require a unique energy control procedure to deal with the servicing and maintenance hazards.
>
> OSHA will carefully examine the energy control procedure of any employer who claims that only one comprehensive procedure is necessary in its workplace to ensure that the single procedure is indeed adequate. While the agency does not insist on a multiplication of procedures, it does insist that they not become so complex that service and maintenance employees cannot easily follow them. In order to qualify as one procedure, the supplemental checklists or appendices must support the basic comprehensive (generic) energy control procedures by having the same:
>
> (1) intended uses for different machines or equipment;
>
> (2) procedural steps for shutting down, isolating, blocking and securing machines or equipment to control hazardous energy;
>
> (3) procedural steps for the placement, removal and transfer of lockout or tagout devices and the responsibility for them; and
>
> (4) requirements for testing a machine or equipment to determine and verify the effectiveness of lockout devices, tagout devices, and other energy control measures."

Where a generic or common procedure is used, it must provide a means for the authorized employees to determine its applicability to the particular machines or pieces of equipment. In this regard, OSHA Instruction STD 1-7.3 provides that the common procedure must identify the equipment "at least by type and location." However, OSHA further advises in a June 20, 1996 letter of interpretation that "[a]ny method of identification which ensures that an authorized employee can determine which energy control procedure applies to the particular machine or group of machines on which servicing and maintenance is to be performed is acceptable to OSHA. When the same energy control procedure can be used to perform servicing and maintenance on all the same type machines or equipment in a workplace, those machines or equipment may be identified by type and model number rather than by type and location. In other situations, however, identification by type and location may be necessary, since energy control procedures for machines or equipment of a particular type may differ based on their application or how they are installed."

ECP Modification

Occasionally, changes to the ECPs may be needed to address deficiencies or to incorporate improvements. For example, the *Periodic Inspection* provisions of paragraph (c)(6) of the standard require that ECPs be reviewed at least annually to ensure that they are adequate, updated, and being followed. According to OSHA, *each* ECP must be separately inspected. Any deficiencies or inadequacies in the ECPs identified during the periodic inspection must be corrected by modifying the ECP as appropriate.

Another situation that warrants ECP modification is an accident investigation that reveals that the ECP itself was not adequate. Inadequacies could be attributed to errors in the procedure, or a problem with the procedure being too general, or not otherwise handling a specific application effectively. Other circumstances that warrant modification of an ECP include the introduction of new (or different) methods of energy control, changes in the location of isolating devices, energy sources, or involved employees, or other changes in the equipment that might require different procedural steps. When necessary, ECPs should be modified to reflect these types of changes.

Figure 8.1. Energy Control Procedure Form and Instructions

ABC Corporation ENERGY CONTROL PROCEDURE				
Equipment Description:[1]				
Equipment Number:[2]				
Job(s):[3]				
Hazardous Energy		**Isolating Devices**		
Type[4]	Magnitude[5]	Type[6]	Location[7]	ID Number[8]
Involved Personnel				
Authorized Employees[9]			Affected Employees[10]	

✔[11]	**Step #**	**Procedure[12]**
	1	
	2	
	3	
	4	
	5	use additional sheets if necessary

Preparer's name:[13] Date:[14]

Figure 8.1. Energy Control Procedure Form and Instructions (Continued)

ABC Corporation
ENERGY CONTROL PROCEDURE

Instructions for completing the Energy Control Procedure Form:

1. **Equipment Description.** Describe the machine, equipment, or process to be locked/tagged out prior to service or maintenance.
2. **Equipment ID Number.** List the unique identification number of the machine, equipment, or process to be locked/tagged out.
3. **Job(s).** Describe the intended use of the procedure or the type of service and/or maintenance to be performed. Service or maintenance includes activities such as constructing, installing, setting up, adjusting, inspecting, modifying, and maintaining and/or servicing machines or equipment and making adjustments or tool changes, where the employees may be exposed to the unexpected energization or startup of the equipment or release of hazardous energy.
4. **Type of Hazardous Energy.** List all types of hazardous energy associated with the machine, equipment, or process to be controlled, including electrical, mechanical, hydraulic, pneumatic, chemical, thermal, or other energy.
5. **Magnitude of Hazardous Energy.** Quantify the energy source to be controlled (e.g. volts, PSI, degrees Fahrenheit, etc.)
6. **Type of Isolating Device.** Identify the mechanical device that physically prevents the transmission or release of energy. Examples of isolating devices include: a manually operated electrical circuit breaker; a disconnect switch; a manually operated switch by which the conductors of a circuit can be disconnected from all ungrounded supply conductors and, in addition, no pole can be operated independently; a slide gate; a line valve; a block; and any similar device used to block or isolate energy.
7. **Location of Isolating Device.** List the location of the isolation device as specifically as possible (e.g. "2nd floor switchgear room" or "machine base opposite flywheel").
8. **ID Number for Isolating Device.** List the unique identification number of the isolating device if available.
9. **Authorized Employees.** List all authorized employees who will be involved with the service or maintenance and their trade. An authorized employee is defined as "a person who locks out or tags out machines, equipment, systems, or system components to perform service or maintenance."
10. **Affected Employees.** List all affected employees. An affected employee is defined as "an employee whose job requires him/her to operate or use a machine or equipment on which servicing or maintenance is being performed under lockout or tagout, or whose job requires him/her to work in an area in which such servicing or maintenance is being performed"
11. ✓. Place a check mark in the box beside each step of the procedure as completed.
12. **Procedure.** List the specific procedures and sequences for de-energization, lockout/tagout application, verification, and restoration to normal operation for the machine, equipment, or process. The person or position responsible for each step which isolate equipment or energy should also be listed. Procedures must be modified as necessary to reflect procedural and/or equipment changes.
13. **Preparer's Name.** The person who prepared the procedure should sign his/her name in ink in the space provided.
14. **Date.** List the date on which the procedure was written. Procedures must be reviewed at least annually.

Figure 8.2. Energy Control Procedure for 5-Gallon Bottle Packer

<table>
<tr><th colspan="6">ABC Corporation
ENERGY CONTROL PROCEDURE</th></tr>
<tr><td colspan="2">Equipment Description:</td><td colspan="4">5-Gallon Bottle Packer</td></tr>
<tr><td colspan="2">Job(s):</td><td colspan="4">Inspection, lubrication, adjustment, repair, or replacement of power transmission components (e.g. belts, chains, etc.). conveyor, drive motor, rotors, rollers, or other auxiliary parts; clearing of bottle jam when employee places any part of body into the danger zone; or other service or maintenance task where employees may be exposed to the hazardous energy types below.</td></tr>
<tr><th colspan="2">Hazardous Energy</th><th colspan="4">Isolating Devices</th></tr>
<tr><td>Type</td><td>Magnitude</td><td>Type</td><td colspan="2">Location</td><td>ID Number</td></tr>
<tr><td>electrical</td><td>480 V</td><td>disconnect</td><td colspan="2">packer control panel</td><td>n/a</td></tr>
<tr><td>pneumatic</td><td>80 psi</td><td>ball valve</td><td colspan="2">on column adjacent to bottle scoop</td><td>n/a</td></tr>
<tr><td>potential</td><td>450 lbs.
(elevator rack)</td><td>block/brace</td><td colspan="2">secured beneath elevator rack</td><td>n/a</td></tr>
<tr><th colspan="6">Involved Personnel</th></tr>
<tr><td colspan="3">Authorized Employees</td><td colspan="3">Affected Employees</td></tr>
<tr><td colspan="3">list</td><td colspan="3">list</td></tr>
<tr><th>✔</th><th>Step #</th><th colspan="4">Procedure
To be used in conjunction with Ionics' Policy for Lockout/Tagout–the Control of Hazardous Energy. If any of the following steps cannot be completed, or if after completing the procedure de-energization cannot be confirmed, DO NOT PROCEED with service or maintenance and contact your supervisor.</th></tr>
<tr><td></td><td>1</td><td colspan="4">Shut machine down by normal stopping means. At the operator control station, place the "manual, off, auto" switch to the "manual" position, and place the keyed control switch to the "off" position.</td></tr>
<tr><td></td><td>2</td><td colspan="4">Open the main disconnect and attach lock(s) and tag(s).</td></tr>
<tr><td></td><td>3</td><td colspan="4">Close the ball valve for the compressed air line and attach lock(s) and tag(s). Bleed residual pressure from the line by opening the bleed screw at the base of the air line pressure regulator. Ensure that residual air line pressure has been bled off by verifying that the regulator pressure gauge reads "0" PSI.</td></tr>
<tr><td></td><td>4</td><td colspan="4">If necessary, lower the elevator to its resting position. If this is not possible and work must be performed with the elevator in a raised position, secure the block designed for this purpose in place to prevent it from falling.</td></tr>
</table>

Figure 8.2. Energy Control Procedure for 5-Gallon Bottle Packer (Continued)

ABC Corporation ENERGY CONTROL PROCEDURE		
✔	Step #	Procedure
	5	Verify isolation–attempt to start the packer by pushing the “Control Power On” switch. Caution: return controls to the “off” position before proceeding. If work is to be performed on electrical circuits or components (e.g. motors, feeder, and/or control circuits, etc.), confirm de-energization of all elements to which employees may be exposed by using appropriate electrical test equipment.
	6	Perform service or maintenance.
	7	Restoration of equipment to normal operation–verify equipment is in a condition to be safely energized. Verify that all non-essential items (e.g. tools, blocks, grounds, etc.) have been removed, that the equipment is operationally intact, and that controls are in a safe (i.e., “off”) position.
	8	Verify that all persons are clear of circuits and equipment–before circuits and/or equipment are re-energized, visually verify that all persons are in the clear, and notify area personnel that the equipment will be reenergized.
	9	Remove locks and tags from the main disconnect and compressed air line ball valve.
	10	Re-energize equipment–close the main disconnect, open the air line ball valve, and return system to normal operation per system operating procedures.
	11	End of procedure.

Preparer's name: I.M. Safer Date: 12/1/00

Figure 8.3. Energy Control Procedure for Palletizer

ABC Corporation ENERGY CONTROL PROCEDURE	
Equipment Description:	Palletizer
Equipment ID Number:	n/a
Job(s):	Inspection, lubrication, adjustment, repair, or replacement of power transmission components (chains, air cylinders, etc.), drive motors, rollers; clearing case jams or damaged cases when employee places any part of body into a danger zone; or other service or maintenance task where employees may be exposed to the hazardous energy types below.

Hazardous Energy		**Isolating Devices**		
Type	Magnitude	Type	Location	ID Number
electrical	480 VAC	disconnect	palletizer control panel	n/a
pneumatic	80 psig	ball valve	back of machine	n/a
potential	200 lbs. (weight of hoist)	locking pins	back of machine	n/a

Involved Personnel	
Authorized Employees	Affected Employees
Forklift operators, shift supervisors, maintenance personnel	Forklift operators, other area personnel

✔	Step #	Procedure
	1	When possible, lower the hoist to its resting position. Otherwise, hoist must be secured in place with locking pins.
	2	Shut machine down by normal stopping means. At the control panel, place the “on/off” switch in the “off” position and the infeed “‘on/off” switch in the “off” postition.
	3	Shut off main power to the control panel and attach lock(s) and tag(s).
	4	Close the ball valve for the compressed air line and attach lock(s) and tag(s). Bleed residual pressure from the line by opening the bleed screw at the base of the air line pressure regulator.

Figure 8.3. Energy Control Procedure for Palletizer (Continued)

ABC Corporation ENERGY CONTROL PROCEDURE		
✔	Step #	Procedure
	5	Verify isolation: a. Attempt to start the palletizer by pressing the green "start" button on the control panel. b. If work is to be performed on electrical circuits or components, confirm de-energization of all elements to which employees may be exposed by using appropriate electrical test equipment. c. Verify that the regulator pressure gauge reads "0" PSI.
	6	Perform service or maintenance.
	7	To restore equipment back to normal operation, remove locking pins and: a. Verify equipment is in a condition to be safely energized. b. Verify that all non-essential items (e.g. tools, etc.) have been removed, that all guards have been replaced, air pressure bleed screws are closed, and that the equipment is operationally intact. c. Check that all controls are in the "off" position.
	8	Verify that all personnel are clear of circuits and equipment before power is restored, and notify all authorized and affected persons that the equipment will be re-energized.
	9	Remove locks and tags from main disconnect and compressed air line ball valve.
	10	Re-energize equipment: a. Close the main disconnect. b. Open the air line ball valve. c. Return system to normal operation per system operating procedures.
	11	End of procedure.

Preparer's name: I.M. Safer — Date: 12/1/00

Figure 8.4. Energy Control Procedure for Rotary Die Cutter

<table>
<tr><th colspan="6">ABC Corporation
ENERGY CONTROL PROCEDURE</th></tr>
<tr><td colspan="2">Equipment Description:</td><td colspan="4">Rotary Die Cutter</td></tr>
<tr><td colspan="2">Equipment ID Number:</td><td colspan="4">n/a</td></tr>
<tr><td colspan="2">Job(s):</td><td colspan="4">Changing die, installing or removing rolls of material, other service or maintenance tasks where employees may be exposed to the hazardous energy sources below.</td></tr>
<tr><th colspan="2">Hazardous Energy</th><th colspan="4">Isolating Devices</th></tr>
<tr><td>Type</td><td>Magnitude</td><td>Type</td><td colspan="2">Location</td><td>ID Number</td></tr>
<tr><td>electrical</td><td>600 V</td><td>circuit breaker</td><td colspan="2">On wall to the right of the rotary die cutter.</td><td>n/a</td></tr>
<tr><td>electrical</td><td>600 V</td><td>local disconnect</td><td colspan="2">main control panel on left side of machine.</td><td>n/a</td></tr>
<tr><td>pneumatic</td><td>80 PSI</td><td>2-way air valve</td><td colspan="2">on air line at left end of equipment</td><td>n/a</td></tr>
<tr><th colspan="6">Involved Personnel</th></tr>
<tr><td colspan="3">Authorized Employees</td><td colspan="3">Affected Employees</td></tr>
<tr><td colspan="3">Qualified rotary die operators, qualified maintenance workers</td><td colspan="3">Rotary die workers; spacer glue machine operator, other area personnel</td></tr>
<tr><th>✔</th><th>Step #</th><th colspan="4">Procedure</th></tr>
<tr><td></td><td>1</td><td colspan="4">Shut down the rotary die cutter by normal stopping means–three sections of the machine will need to be de-energized and isolated:
1. UNWIND–shut down unwind motor by moving POWER switch on MAG-POWR panel to OFF position. Panel is located on left side of unit. On right side of unit, move switch on electrical box to the OFF position.
2. DIE CUTTER–switch Baldor DC drive panel switch (located on right side of equipment) to OFF position. On Control Panel 1, press red STOP button. On Control Panel 2, press red STOP button labeled REWIND, move switch label UNWIND to OFF position, and press red STOP button labeled "Trim Removal."
3. REWIND–To shut down rewind motor, move POWER switch on MAGPOWR panel to OFF position. Panel is located on left side of unit. Push POWER switch to OFF position on both MIZER PLUS panels (there are 2 panels).</td></tr>
<tr><td></td><td>2</td><td colspan="4">Open the local disconnect (move to OFF position) on the main control panel located on the left side of the machine.</td></tr>
</table>

Figure 8.4. Energy Control Procedure for Rotary Die Cutter (Continued)

ABC Corporation ENERGY CONTROL PROCEDURE		
✔	Step #	Procedure
	3	Close the air supply valve (move to OFF position); the handle should be in the horizontal position. Bleed residual pressure from the air line by opening the bleed screw at the base of the air line pressure regulator.
	4	Open the main circuit breaker located on the wall to the right side of unit. All authorized employees place locks and tags on breaker isolating device to secure it in the open position.
	5	Verify isolation. Confirm 0 PSI at pressure gauge in-line with 2-way air valve. Confirm that rotary die cutter cannot be energized; attempt to start the die cutter by moving the UNWIND motor power switch start to the ON position. Caution: return controls to the OFF position before proceeding. Confirm that all rotating dies and rollers have come to a complete stop.
	6	Perform above listed service or maintenance job(s) per manufacturer's instructions.
	7	Restoration of equipment to normal operation. Verify that equipment is in a condition to be safely energized. Verify that all non-essential items (e.g. tools, etc.) have been removed, that the equipment is operationally intact, air pressure bleed screws are closed, and that switches/controls are in a safe or off position.
	8	Verify that all persons are clear of circuits and equipment. Before equipment is re-energized, visually verify that all persons are in the clear, and notify all authorized and affected persons that the equipment will be re-energized.
	9	Remove lock(s) and tag(s) from main circuit breaker.
	10	Re-energize equipment. Close main circuit breaker and local disconnect. Open 2-way air valve and return system to normal operation per manufacturer's instructions.
	11	End of procedure.

Preparer's name: I.M. Safer Date: 12/1/00

Figure 8.5. Energy Control Procedure for Flow International Waternife 20X W

ABC Corporation ENERGY CONTROL PROCEDURE				
Equipment Description:		Flow International Waternife 20XW		
Equipment ID Number:		G1101MS		
Job(s):		Tip change-out, Garnet metering disc change-out		
Hazardous Energy		**Isolating Devices**		
Type	Magnitude	Type	Location	ID Number
electrical	480 V	local disconnect	On wall behind and left of pump controller	n/a
hydraulic (high-pressure water)	55,000 PSI	ball valve	rear of pump controller	n/a
hydraulic (high-pressure water)	55,000 PSI	ball valve	right rear of gantry	n/a

Involved Personnel	
Authorized Employees	Affected Employees
List	List

✔	**Step #**	**Procedure**
	1	Shut down Waternife by normal stopping means–press stop button on pump controller panel keypad.
	2	Open local disconnect marked "Waternife" on wall behind and left of pump controller, and attach lock(s) and tag(s)
	3	Ensure both valves on high pressure water line are OPEN (pumps and line are self-bleeding), and verify no pressure in line by confirming 0 PSI at pressure gauge at right rear of gantry.
	4	Confirm that Waternife can not be energized by attempting to start by pressing start button on pump controller or should this be done at the MTB panel. Caution: Return controls to the "off" position before proceeding.
	5	Perform above listed service or maintenance job(s) per manufacturer's instructions.

Figure 8.5. Energy Control Procedure for Flow International Waternife 20XW (Continued)

ABC Corporation ENERGY CONTROL PROCEDURE		
✔	Step #	Procedure
	6	Restoration of equipment to normal operation. Verify that equipment is in a condition to be safely energized. Verify that all non-essential items (e.g. tools, etc.) have been removed, that the equipment is operationally intact, and that controls are in a neutral/safe position.
	7	Verify that all persons are clear of circuits and equipment. Before equipment is re-energized, visually verify that all persons are in the clear, and notify all authorized and affected persons that the equipment will be re-energized.
	8	Remove lock(s) and tag(s) from local disconnect.
	9	Re-energize equipment. Close local disconnect and return system to normal operation per manufacturer's instructions.
	10	End of procedure.

Preparer's name: I.M. Safer Date: 12/1/00

Chapter 9

Employee Training

The training and education of employees in the recognition, avoidance, and/or control of energy-related hazards is the cornerstone of an effective lockout/tagout program; without them, the viability of the program can be seriously compromised. In the 1999 NIOSH Publication *A Model for Research on Training Effectiveness*, "training" is defined as "communication directed at a defined population for the purpose of developing skills, modifying behavior, and increasing competence. Generally, training focuses exclusively on *what* needs to be known." "Education", on the other hand, is characterized as "a longer-term process that incorporates the goals of training and explains *why* certain information must be known." Both training *and* education are needed to provide employees with the knowledge, skills, and motivation needed to effectively launch and maintain a successful lockout/tagout program.

While the effects of training employees cannot be exactly quantified, it is generally believed that well-trained employees are far less likely to be involved in accidents than their untrained counterparts. A 1998 NIOSH study examining the merits of training (*Assessing Occupational Safety and Health Training,* DHHS/NIOSH Publication No. 98-145), suggests the benefits of training to overall job safety, and implicates the lack of training as a contributing factor to mishaps. A 1980 Bureau of Labor Statistics (BLS) Work Injury Report Study also suggests an adverse effect associated with a *lack* of training on energy control measures (see Table 9.1 below). Out of a total of 613 injured workers surveyed during the study, more than half had received no training on lockout/tagout.

Table 9.1: Lockout Training of Injured Employees.

Type of Training	Number	Percent
Printed instruction	25	4
Procedures posted on equipment	37	6
Training at job orientation, at meetings, or otherwise	211	34
No training	340	55

Source: BLS Work Injury Report Study (from 613 responses)

Of those injured employees who had received training, 15 stated that their training had occurred *after* their accident. Additionally, 60 employees stated that they had received their training more than a year prior to the accident. Even if training had been provided at some time during employment, the length of time between the receipt of the training and the accident appears to have limited any benefit derived from the training. In promulgating the lockout/tagout standard, therefore, OSHA recognized the need for remedial or refresher training of employees who must use lockout/tagout, and incorporated provisions for periodic retraining of employees and for annual inspection of energy control procedures.

In the publication *Training Requirements in OSHA Standards and Training Guidelines* (OSHA Publication No. 2254), OSHA outlines voluntary guidelines to assist employers in providing safety and health training of their employees. Seven steps are recommended for conducting, evaluating, and revising training programs:

1. Determine if training is needed;
2. Identify training needs;
3. Identify goals and objectives;
4. Develop learning activities;
5. Conduct the training;
6. Evaluate program effectiveness; and
7. Revise/improve the program.

The guidelines also recommend that training be conducted:

- For new and reassigned employees;
- When new equipment or processes are introduced;
- When procedures have been revised or updated;
- When new information must be made available; and
- When employee performance needs to be improved.

Following these guidelines will help employers launch and maintain an effective lockout/tagout training program.

A general assumption about human behavior is that "people will do what's expected of them." Training, in its many forms, is a key mechanism for communicating to employees what's expected of them. Proper training on energy control procedures and an understanding of the reasons for the energy control program are crucial to their implementation by the employees. Training and retraining on lockout/tagout is needed to ensure the continued proficiency of employees in following safe work practices. A trait shared by companies with successful lockout/tagout programs is that they have developed and implemented extensive training programs to ensure that their employees acquire and retain the necessary knowledge and skills. Successful training is also influenced by other factors, including the role and support of management. Monitoring work practices and providing feedback to employees after the training are needed to reinforce the training and motivate employees to use the knowledge gained.

Initial Training Requirements

Requirements for employee training on the energy control program are given in section (c)(7) of the standard:

> 1910.147 (c)(7)
>
> *Training and communication.* The employer shall provide training to ensure that the purpose and function of the energy control program is understood by employees and that the knowledge and skills required for the safe application, usage and removal of the energy controls are acquired by employees.

OSHA Instruction Standard 1-7.3 provides interpretive guidance for inspection and enforcement of the lockout/tagout standard. With respect to employee training, it states in part: "[t]he employer's training program must cover, at a minimum, the following three areas: energy control program, elements of energy control procedures relevant to employee duties, and the pertinent requirements of the standard... The employer must provide: (1) effective initial training; (2) effective retraining as needed; and (3) certification of training."

The level and type of training that each employee must receive depends upon a number of factors, including the:

- relationship of the employee's job to the machine or equipment being locked or tagged out;
- role the employee plays in the control of energy;
- complexity of the equipment and the degree of knowledge about the hazardous energy that he or she must possess to accomplish the assigned tasks safely; and
- procedures used to secure energy isolating devices (that is, whether lockout or a tagout system alone is used).

For the purposes of the standard, three types of employees that require training are recognized — *authorized, affected,* and *other.* The relative degree of knowledge required by these three employee types is in descending order, with the requirements for authorized employees being the most stringent. As a basic requirement, all employees should be trained in the basics of lockout/tagout. Through this training, all employees should understand the purpose (i.e., to protect service or maintenance employees from injury associated with the accidental startup of the equipment or release of stored energy), function, and restrictions of the energy control program. The training needed by each employee should be provided initially upon assignment.

An *authorized employee*, by definition, is "a person who locks out or tags out machines or equipment in order to perform servicing or maintenance on that machine or equipment." Authorized employees are those who are charged with the responsibility for implementing the energy control procedures prior to performing the servicing or maintenance, and thus require the highest level of training. In addition to understanding the purpose and function of the energy control program, authorized employees must also receive specific training as specified in paragraph (c)(7)(i)(A) of the standard:

> 1910.147 (c)(7)(i)(A)
>
> Each authorized employee shall receive training in the recognition of applicable hazardous energy sources, the type and magnitude of the energy available in the workplace, and the methods and means necessary for energy isolation and control.

An *affected employee*, by definition, is "an employee whose job requires him/her to operate or use a machine or equipment on which servicing or maintenance is being performed under lockout or tagout, or whose job requires him/her to work in an area in which such servicing or maintenance is being performed." Affected employees are generally the machine operators or users. An affected employee becomes an authorized employee when that employee's duties include performing servicing or maintenance covered under this section.

Training for affected employees is specified in paragraph (c)(7)(i)(B) of the standard:

> 1910.147 (c)(7)(i)(B)
>
> Each affected employee shall be instructed in the purpose and use of the energy control procedure.

With respect to "*other*" employees, this term is not found in the definitions section of the standard. It is referenced in paragraph (c)(7)(i)(C) of the training section of the standard, however:

> 1910.147 (c)(7)(i)(C)
>
> All other employees whose work operations are or may be in an area where energy control procedures may be utilized, shall be instructed about the procedure, and about the prohibition relating to attempts to restart or reenergize machines or equipment which are locked out or tagged out.

Because affected and other employees are not the individuals performing the servicing or maintenance or using the energy control procedures, the scope of their training is less than that of authorized employees. Since they work in areas where equipment is being serviced by authorized employees, however, affected and other employees are potentially in a position to take action that could put the authorized employee at risk (for instance, attempting to restart or reenergize equipment). In addition to understanding the purpose and function of the energy control program, therefore, affected and other employees also need to be able recognize when the energy control procedure is being used and understand the restrictions of the program. They must be instructed not to attempt to start up or use the equipment that has been locked or tagged out. The consequences (disciplinary action) for attempting to start-up locked/tagged out equipment or removing locks or tags without authorization should also be clearly communicated to all employees.

For employees who exclusively perform functions related to normal production operations, and who perform servicing and/or maintenance under the protection of normal machine safeguarding, OSHA advises that these individuals need only be trained as affected rather than authorized employees, even if tagout procedures are used (see *Additional Training Requirements Where Tagout Alone is Used*—below).

Additional Training Requirements Where Tagout Alone is Used

Use of tags alone to secure energy isolating devices on equipment that is not capable of being locked out, although permitted under the standard, does not provide the same assurance as lockout that the equipment won't be accidentally reenergized. In contrast to locks, for example, tags do not actually secure the energy isolating device from movement, and are more easily bypassed or removed than locks. In light of the inherent weaknesses of tagout, the OSHA standard imposes additional training requirements for employees who work with tagout, or who work in areas where a tagout system alone is used.

These additional training requirements are specified in section (c)(7)(ii) of the standard:

> 1910.147 (c)(7)(ii)
>
> When tagout systems are used, employees shall also be trained in the following limitations of tags:
>
> (c)(7)(ii)(A)
>
> Tags are essentially warning devices affixed to energy isolating devices, and do not provide the physical restraint on those devices that is provided by a lock.
>
> (c)(7)(ii)(B)
>
> When a tag is attached to an energy isolating means, it is not to be removed without authorization of the authorized person responsible for it, and it is never to be bypassed, ignored, or otherwise defeated.
>
> (c)(7)(ii)(C)
>
> Tags must be legible and understandable by all authorized employees, affected employees, and all other employees whose work operations are or may be in the area, in order to be effective.
>
> (c)(7)(ii)(D)
>
> Tags and their means of attachment must be made of materials which will withstand the environmental conditions encountered in the workplace.

(c)(7)(ii)(E)

Tags may evoke a false sense of security, and their meaning needs to be understood as part of the overall energy control program.

(c)(7)(ii)(F)

Tags must be securely attached to energy isolating devices so that they cannot be inadvertently or accidentally detached during use.

When a tagout system alone is used, the employer must meet more stringent program requirements intended to improve its capability and enhance employee protection. For example, the employer must demonstrate that the tagout system provides "full employee protection." To demonstrate full employee protection, the employer must implement additional safety measures as necessary to achieve a level of safety equivalent to that of a lockout system. Examples of additional measures cited by OSHA include the removal of an isolating circuit element, blocking of a controlling switch, opening of an extra disconnecting device, or the removal of a valve handle to reduce the likelihood of inadvertent energization. The equipment-specific energy control procedures must set forth the steps for placement, removal, and transfer of tags, and for implementing any additional measures that are necessary. Training should be provided to ensure that employees understand and follow the steps that are prescribed.

Retraining

It is widely believed that the effectiveness of training diminishes as the time from the last training session increases. Retraining, therefore, is a necessary element for maintaining an effective energy control program, and is a way of keeping lockout/tagout practices and procedures fresh in the mind of employees. Along with supervision, enforcement, and periodic inspections, retraining is a means for ensuring consistent implementation of lockout/tagout procedures, and the continued proficiency of employees in the safe application of energy controls. Retraining is also valuable as a tool for addressing unsafe behaviors reinforcing safe behaviors, and as an intervention effort to prevent accident recurrence.

Requirements for employee retraining are specified in section (c)(7)(iii) of the standard:

1910.147(c)(7)(iii). Employee retraining.

(c)(7)(iii)(A)

Retraining shall be provided for all authorized and affected employees whenever there is a change in their job assignments, a change in machines, equipment or processes that present a new hazard, or when there is a change in the energy control procedures.

(c)(7)(iii)(B)

Additional retraining shall also be conducted whenever a periodic inspection under paragraph (c)(6) of this section reveals, or whenever the employer has reason to believe that there are deviations from or inadequacies in the employee's knowledge or use of the energy control procedures.

(c)(7)(iii)(C)

The retraining shall reestablish employee proficiency and introduce new or revised control methods and procedures, as necessary.

Clearly, the provisions of (c)(7)(iii)(A) are intended to remedy any potential shortcomings of previous training that result from changes in job assignments, equipment, or energy control procedures. The language of (c)(7)(iii)(B), however, is intended to address problems associated with employee performance. For example, retraining is required if an audit, inspection, observation, or investigation reveals deviation from established procedures or inadequacies in an employee's knowledge, abilities, or use of the energy control procedures. In this regard, section (c)(6) of the lockout/tagout standard requires employers to conduct a periodic inspection of the energy control procedure at least annually to ensure that the procedure and the requirements of the standard are being followed, and to correct any deviations or inadequacies identified. Retraining provides the means for correcting employee-related issues that arise.

Other events that could trigger retraining are described in the preamble to the lockout/tagout Final Rule. "For example, an employee working with an energy control procedure might be injured in the course of his duties, or there might be a 'near miss,' where no one is actually injured, but where the energy control program has failed nonetheless. If a subsequent investigation indicates that an employee failed to operate within the guidelines of the control procedure, retraining would be required. In addition, the investigation might also reveal that the procedure itself was not adequate. Such inadequacies in the procedure could be the result of using a general procedure that does not handle effectively a specific application, or they may arise because changes have been made to the equipment or process that were not taken into account by the existing energy control procedure. In such cases when changes to the energy control procedure must be made, the employer is required to retrain employees in the new or revised procedure...." When changes are made to the energy control procedure, therefore, retraining should account for any differences in the machines or equipment or how they are used, and in the energy control procedures that apply to them. If warranted by the severity of the problem encountered with the use of the energy control program/procedure, the retraining should be more concentrated or encompassing than the initial training.

Delivery of Training

Rather than stipulating how training should be accomplished, the language of the lockout/tagout standard is performance based, and focuses on *what* needs to be known by employees. For example, all employees must "understand the purpose and function of the energy control program." Authorized employees must be trained "in the recognition of applicable hazardous energy sources, the type and magnitude of the energy available in the workplace, and the methods and means necessary for energy isolation and control." Other employees must be instructed "about the [energy control] procedure," and "about the prohibition relating to attempts to restart or reenergize machines or equipment which are locked out or tagged out." The degree of training provided should be determined by the risk to the employee. As far as how the training is to be conducted, however, employers are provided with considerable flexibility, and the employer is permitted to use whatever method he/she feels will best accomplish the objectives of the training.

A number of variables should be considered in establishing a lockout/tagout training program. The knowledge and skill levels of employees, the size of the training group, the length and/or frequency of training, the manner of instruction, and trainer credentials are all factors to be considered in how best to deliver the training. For example, training can be group oriented, or conducted one-on-one. Various training methods and media can also be used with varying degrees of success, including classroom instruction, computer-based instruction, lectures, safety meetings, demonstrations, handouts, videos, supervision and on-the-job training. PowerPoint presentations, slides, overheads, videos, flow charts, manuals, and so on are all useful tools available to the trainer for making high-impact presentations. In many cases, a combination of classroom, multi-media, and hands-on training is most effective.

The use of computer-based training, videos, or simple handouts alone has limitations, and is generally not suggested. Training is more effective when employees become involved in the training process by participating in discussions, asking questions, contributing knowledge and expertise, and learning through hands-on experiences, role-playing exercises, and similar activities. If a company lacks the necessary in-house expertise to conduct the training, the use of external resources such as local safety councils, consultants, associations, and academia should be considered. OSHA-funded state onsite consultation programs, and courses offered through OSHA's Office of Training and Education are also resources to be considered.

An excellent resource is the *Lockout/Tagout Interactive Training Program* developed by OSHA's Directorate of Technical Support, available for review on OSHA's website at www.osha.gov. This interactive, computer-based training program is a web application designed and developed to provide the viewer with an in-depth understanding of the lockout/tagout standard. The program includes: a tutorial that summarizes the key components of the standard in a question/answer format; a detailed discussion of five "hot topic" areas; and interactive case studies based on descriptions of lockout/tagout inspections derived from compliance interpretations, court decisions, Review Commission decisions, and inspection files. Simulated lockout/tagout inspections are presented, and the participant is prompted to make decisions on the application of the standard based on information presented on the screen.

Basic Training

To meet the training requirements of the standard, basic training on lockout/tagout should be provided to *all* employees. The basic training should explain the purpose, function, and restrictions of the energy control program, and the consequences for non-compliance.

Figure 9.1. Effective lockout/tagout training starts in the classroom

This basic training can be accomplished in a number of ways, including new employee orientation sessions, computer-based instruction, the use of employee handbooks, and through scheduled safety meetings, Many employers distribute the company lockout/tagout policy to employees and require them to sign a form indicating that they have received and read the policy and understand its requirements.

It is also a good idea to provide training on other elements in addition to the minimum requirements prescribed in the standard. For example, it is also useful for all employees to have at least a basic understanding of the types of hazardous energy used by the machines, equipment, and processes in their workplace. Reviewing accident and injury records and/or case studies to educate employees on how accidents occur is also beneficial. Under this approach, the training requirements for both affected and other employees can be met. For convenience, many employers find it useful to train affected and other employees together.

Authorized employees, as the individuals with responsibility for implementing the energy control procedures and for performing the service or maintenance tasks, must receive more extensive training than affected and other employees. In addition to receiving basic training on the purpose, function, and restrictions of the energy control program (as described above for affected and other employees), authorized employees must also receive training in the recognition of applicable hazardous energy sources, the type and magnitude of the energy available in the workplace, and the methods and means necessary for energy isolation and control. Specific instruction should also be provided on the detailed energy control procedures and any unique hazards associated with the particular equipment to be serviced or maintained.

In most cases, a combination of classroom instruction and hands-on field training is most effective in ensuring that authorized employees have the knowledge and skills necessary for the safe application, use, and removal of energy controls. Classroom instruction, for example, is useful for covering the requirements of the OSHA standard, the hazardous energy types and magnitudes to be encountered, basic concepts and methods of energy control, and the policies and procedures of the employer's energy control program (see Figure 9.1). The general procedures for equipment shutdown and isolation, stored energy release or restraint, application and removal of locks, tags, and other devices, verification of de-energization, and release from lockout/tagout (returning equipment back to service) can also be covered during the classroom portion of the training.

On-the-Job Training

On-the-job training that supplements the basic classroom instruction is also needed for authorized employees. This training should be specific to the particular equipment to be serviced or maintained, and should cover the actual methods, techniques, and devices that are used to control or isolate hazardous energy. Thorough instruction and hands-on practice in the use and operation of equipment, including energy isolating devices, protective materials and hardware, and test instrumentation is needed to ensure proficiency.

The familiarization of authorized employees with the equipment to be serviced is also critical. In this regard, apprenticeship and other programs where employees work under the supervision of an experienced employee are invaluable for transferring needed skills and knowledge to new employees. If necessary, the applicable provisions of the equipment manufacturer's operations and maintenance manual should be reviewed by involved employees as part of their training. For employees who both operate and service machines, on-the-job lockout/tagout training should ideally be integrated with other aspects of job training.

Energy Control Procedures

Along with the written energy control program that defines the purpose, scope, application, and responsibilities of lockout/tagout, employers are required to develop more detailed, equipment-specific energy control procedures (ECP). The ECPs supplement and work in conjunction with the general provisions of the overall written program, and define the *specific* steps to be taken to de-energize and isolate machines, equipment, or processes. The establishment of ECPs must be coupled with sufficient on-the-job training for authorized employees to ensure their successful use.

Training for Supervisors and Managers

Supervisors and managers are also key individuals with leadership and oversight roles in the lockout/tagout program. In addition to ensuring that employees are properly trained, important responsibilities of supervisors and managers also include:

- surveying their work areas to identify the machines, equipment, and processes (MEP) that use hazardous energy and that are covered by the program;
- ensuring that equipment-specific energy control procedures are established for the machines, equipment, and processes in their areas;
- ensuring that employees and contractors follow established lockout/tagout and energy control procedures;
- providing protective equipment, hardware and appliances (for example, locks, lockout devices, keys, and tags) to employees;
- maintaining exclusive control of duplicate or master keys;
- ensuring that periodic inspections are conducted and documented as required.

Carrying out these responsibilities is at the very core of preventing hazardous energy-related incidents. To ensure that supervisors and managers understand their responsibilities and can effectively oversee the lockout/tagout program, therefore, more thorough and detailed training for them is often conducted.

Certification of Training

Paragraph (c)(7)(iv) of the standard requires that employers certify that the necessary training has been conducted. The certification must contain each employee's name and the date of training. Records of both the initial training and any periodic retraining must be kept.

> 1910.147 (c)(7)(iv)
>
> The employer shall certify that employee training has been accomplished and is being kept up to date. The certification shall contain each employee's name and dates of training.

The certification required by paragraph (c)(7)(iv) does not provide an indication of the quality or content of the training; it simply shows that training has been performed. In addition to the certification of training, therefore, employers should also be able to demonstrate that employees have been *adequately* trained. In this regard, the adequacy of the training can be evaluated by comparing the elements of the training to the duties and responsibilities of employees under the energy control program and procedures. To document the quality and content of the training provided, it is advisable to keep a program of instruction on file that provides a record of all the information and material that was presented during the training. Records that show the duration and location of the training as well as the trainer's credentials should also be kept.

Inspection guidelines in OSHA Instruction Standard 1-7.3 provide insight on how an OSHA Inspector might evaluate the adequacy of employee training:

> OSHA Instruction Standard 1-7.3
>
> (H)(3)(c). Evaluate the employer's training programs for "authorized", "affected", and "other" employees. Interview a representative sampling of selected employees as a part of this evaluation.
>
> (1) Verify that the training of authorized employees includes:
>
> (a) Recognition of hazardous energy;
>
> (b) Type and magnitude of energy found in the workplace;
>
> (c) The means and methods of isolating and/or controlling energy; and
>
> (d) The means of verification of effective energy control, and the purpose of the procedures to be used.
>
> (2) Verify that affected employees have been instructed in the purpose and use of the energy control procedures.
>
> (3) Verify that all other employees who may be affected by the energy control procedures are instructed about the procedure and the prohibition relating to attempts to restart or reenergize such machines or equipment.

(4) When the employer's procedures permit the use of tagout, the training of authorized, affected, and other employees shall include the provisions of 29 CFR 1910.147(c)(7)(ii) and (d)(4)(iii).

Other Training Considerations

Removal of a Lock or Tag by an Employee Other Than the One Who Applied It

Under the exception noted in paragraph (e)(3) of the standard, when the authorized employee who applied the lockout or tagout device is not available to remove it, that device may be removed by another employee under the direction of the employer, only if the specific procedures and *training* for such removal have been developed, documented, and incorporated into the employer's energy control program. The procedures must include, at a minimum, the following items:

- first, verification that the authorized employee is not at the facility;
- second, making all reasonable efforts to contact that employee to inform him/her that his/her device has been removed; and
- third, ensuring that employee knows of that device removal before he/she resumes work at the facility.

These steps are necessary to ensure that the employee who is protected by the device is not exposed to energy hazards after its removal.

Additional Training for Electrical Work

Additional OSHA training requirements apply to employees who perform electrical work or who are otherwise exposed to electrical hazards. For example, pursuant to the Electrical Safety Related Work Practices provisions of 29 CFR 1910.331–.399, all employees who face a risk of electric shock, burns, or other related injuries (not reduced to a safe level by the installation safety requirements of OSHA's Subpart S regulations) must be trained in and familiarized with the selection and use of safety-related work practices, safety procedures, and other personnel safety requirements that pertain to their respective job assignments. All such employees should be trained in the potential hazards associated with electricity, such as hazardous voltages, electric current, arcing, improper grounding, and lack of guarding.

The level of electrical safety training that needs to be provided depends on the responsibilities of the employee and the degree of risk. *Qualified* employees, for example, are those employees permitted to work on or near *exposed* energized parts. Qualified employees must be "familiar with the construction and operation of the equipment and the hazards involved," and their training must include at the minimum the following:

- the skills and techniques necessary to distinguish exposed live parts from other parts of electric equipment;
- the skills and techniques necessary to determine the nominal voltage of exposed live parts;
- the clearance distances specified in OSHA 1910.333(c) and the corresponding voltages to which the qualified person will be exposed.

Qualified persons whose work on energized equipment involves either direct contact or contact by means of tools or materials must also be "capable of working safely on energized circuits, and must be familiar with the proper use of special precautionary techniques, personal protective equipment, insulating and shielding materials, and insulated tools." In general, qualified employees need to be extensively trained in order for them to perform their work safely.

Unqualified employees are employees who work near energized equipment or circuits that are ***not exposed*** (because of adequate enclosure and so on). These employees should be trained not only to recognize electrical hazards, but also to avoid them. For example, employees should be instructed to be alert at all times when working near electrical conductors, not to contact electrical parts with any body part, tools, or other materials, and not to reach blindly into areas which may contain energized parts. Employees must also be trained in and familiar with any other safety practices not specifically addressed in sections 1910.331 through 1910.335 of the OSHA regulations, but that are related to their work and necessary for their safety.

Use of Electrical Test Instruments and Equipment

In order to verify electrical de-energization during lockout/tagout, employees may be required to use electrical test instruments or equipment. Section 1910.334 of OSHA's *Electrical Safety-Related Work Practices* requires that only *qualified* persons perform electrical testing work, such as taking current and voltage readings or moving wires on electric circuits or equipment.To be considered qualified, employees should first be familiar with the construction and operation of workplace equipment on which the electrical testing will be performed, and the electrical hazards involved. Employees must also be capable of operating and maintaining the electrical test instruments and equipment they use. This capability should be based on formal training on the use of electrical testing instruments, including on-the-job training where the employee demonstrates the ability to safely perform electrical instrument testing.

Chapter 10

Other Lockout/Tagout Requirements and Considerations

Periodic Inspection

The consistent and proper use of energy control procedures is necessary for a lockout/tagout program to maintain its effectiveness. A lapse in implementing energy control procedures, or flaws in the procedures themselves, could potentially subject service or maintenance personnel to serious injury risks.

To ensure that energy control measures in place are adequate and being properly and consistently applied, a lockout/tagout program must be reviewed (inspected) periodically and revised if needed. Section (c)(6) of the OSHA standard sets forth specific requirements for conducting the periodic inspections:

> 1910.147 (c)(6). *Periodic Inspection.*
>
> (c)(6)(i) The employer shall conduct a periodic inspection of the energy control procedure at least annually to ensure that the procedure and the requirements of this standard are being followed.
>
> (A) The periodic inspection shall be performed by an authorized employee other than the one(s) utilizing the energy control procedure being inspected.
>
> (B) The periodic inspection shall be conducted to correct any deviations or inadequacies identified.

The periodic inspection provisions of the standard require that the energy control program be reviewed *at least* annually to ensure compliance. The intent of this requirement is to ensure that energy control procedures are updated, accessible, and being implemented properly, and that employees remain familiar with their responsibilities under those procedures. The periodic inspection also serves as an essential check to verify the continued proficiency of employees in the safe application, use, and removal of the energy controls. The periodic inspection must be designed to correct any deviations or inadequacies observed, and the employer must certify that the periodic inspections have been performed.

Inspection by Authorized Employees

The standard also requires that the inspections be performed by an authorized employee other than one implementing the particular procedure. For purposes of the lockout/tagout standard, OSHA interprets "authorized employee" to mean "a qualified person to whom the authority and responsibility to perform a specific lockout or tagout inspection has been given by the employer." The requirement for the inspector to be an authorized employee is intended to ensure that that person has the necessary knowledge to evaluate the effectiveness of the procedure being inspected. The inspector, for example, should know how the procedures are to be used, and be able to recognize any problems with the energy control program.

Scope of the Inspection

Detailed, equipment-specific energy control procedures must be "developed, documented, and utilized" under the standard [see 1910.147 (c)(4)]. Paragraph (c)(6)(i) of the standard further requires that these energy control procedures be reviewed at least annually as part of the periodic inspection, to ensure that the procedures and the requirements of the standard are being followed. OSHA generally interprets this requirement to mean that *each* energy control procedure must be separately inspected at least annually. Energy control procedures that are used less than once a year are not required to be reviewed annually; these should, however, be reviewed prior to re-use to ensure their adequacy.

The scope of the periodic inspection also depends on whether the company uses a lockout system, or a tagout system alone. Authorized employees are the individuals who implement the energy control procedures, and their safety is a primary consideration. Under a lockout system, it is reasoned that affected employees should not be able to remove locks and place the authorized employees in danger. When a lockout system is in use, therefore, the periodic inspection needs to include *authorized* employees only (including affected employees is not required). The initial training provided to affected employees, where they are instructed that locks are not to be defeated or bypassed, and that locked out equipment must remain de-energized, coupled with the requirement for these employees to be notified before the energy control devices are applied in their work area, is deemed adequate by OSHA.

Where lockout is used, section (c)(6)(i)(C) of the standard requires that the inspector review the responsibilities of each authorized employee under the procedure with that employee.

> 1910.147 (c)(6)(i)(C)
>
> Where lockout is used for energy control, the periodic inspection shall include a review, between the inspector and each authorized employee, of that employee's responsibilities under the energy control procedure being inspected.

In contrast, the standard requires that when tagout alone is used, both authorized *and* affected employees must review the use of the tagout system as part of the annual inspection. This additional review is necessary because of the inherent shortcomings of tagout systems. Tags can be easily removed or bypassed, and provide limited protection when compared to locks. Tags do not actually prevent the activation of the tagged equipment, and the success of tagout relies heavily on the knowledge and training of the employees involved and the continued reinforcement of the meaning of the tags. To ensure proper implementation of tagout, therefore, it is also important to periodically review responsibilities with affected employees and remind them of the restrictions imposed on them under the program. When tagout alone is used, the annual inspection must also include a review on the limitation of tags with each affected and authorized employee.

Section (c)(6)(i)(D) of the standard prescribes the scope of the inspection where tagout systems alone are used:

> 1910.147 (c)(6)(i)(D) Where tagout is used for energy control, the periodic inspection shall include a review, between the inspector and each authorized and affected employee, of that employee's responsibilities under the energy control procedure being inspected, and the elements set forth in paragraph (c)(7)(ii) of this section.

Conducting the Inspection

The preamble to the lockout/tagout Final Rule provides some guidance regarding the inspector's duties in performing a periodic inspection, to assure that necessary information about the energy control procedure and its effectiveness is obtained. The inspection must be able to determine three things:

1. whether the steps in the energy control procedure are being followed;
2. whether the employees involved know their responsibilities under the procedure; and
3. whether the procedure is adequate to provide the necessary protection, and what changes, if any, are needed.

To make these determinations, the inspector needs to observe and talk with involved employees. In this regard, OSHA advises that the inspector does not have to observe *every* authorized employee implementing the energy control procedure on the particular machine or equipment. The inspector will, however, need to:

- observe a representative number of such employees while implementing the procedure; and
- talk with all other authorized employees even though they may not be implementing the energy control procedure.

Potential options for performing the observations include random audits, planned visual observations, or modified plant safety tours.

With respect to the employee reviews that are required to be performed during the periodic inspections by 1910.147 (c)(6)(i)(C) and (D), OSHA suggests that these may be completed in one or more meetings in which all authorized employees (as well as all affected employees when tagout is used) are in attendance to review the specific energy control procedures, as the case may be. One-on-one meetings between the inspector and authorized (and when applicable, affected) employees, although an option, are not necessarily required (see Figure 10.1). Group meetings or training sessions can also be an effective way of accomplishing the periodic inspection reviews, and may be more practical.

Figure 10.1. One-on-one meetings are a good way to conduct periodic inspection reviews.

OSHA further advises that annual refresher training, *by itself*, will not satisfy the periodic inspection review requirements of the standard, since certain benefits to be gained from talks with individual employees may not be addressed in the refresher training. Benefits of these individual talks cited by OSHA include enhanced ability of the inspector to:

- identify inadequacies in, misinterpretations of, and deviations from energy control procedures;
- evaluate task proficiency of authorized employees; and
- provide immediate feedback regarding procedure implementation.

Certifying the Inspection

Under the standard, employers must also certify that the periodic inspections have been performed. Requirements for certifying the periodic inspections are set forth in section (c)(6)(ii) of the standard.

> 1910.147(c)(6)(ii)
>
> (c)(6)(ii) The employer shall certify that the periodic inspections have been performed. The certification shall identify the machine or equipment on which the energy control procedure was being utilized, the date of the inspection, the employees included in the inspection, and the person performing the inspection.

In addition to the equipment, date of inspection, and name of inspector, the employees included in the inspection must also be documented. This requirement is intended to ensure that involved employees have had the opportunity to review their responsibilities and demonstrate their performance under the procedure. An example Periodic Inspection Certification Form is provided in Figure 10.2.

Figure 10.2. Lockout/Tagout Periodic Inspection Certification Form

Lockout/Tagout Periodic Inspection Certification

Facility/Site:	
Department:	
Name/Identification of Machine, Equipment, or Process:	
Energy Control Procedure Being Reviewed:	
Date:	

Name of Employee(s) Being Reviewed (use additional sheets if necessary)		
1.	6.	11.
2.	7.	12.
3.	8.	13.
4.	9.	14.
5.	10.	15.

Inspection items–Review the Energy Control Procedure and employee responsibilities with the involved employees and complete the following:	**Yes**	**No**
1. Are the steps in the energy control procedure being properly followed? (If no, provide a detailed description of the problem below, along with a description of any corrective action taken or planned)		
2. Do the involved employees understand their responsibilities under the procedure? (If no, provide a detailed description of the problem and any corrective action needed below.)		
3. Are there any inadequacies in any employee's knowledge, abilities, or use of the procedures? (If yes, provide a detailed description of the problem and any corrective action needed below).		
4. Is the procedure adequate to provide the necessary protection? (If no, provide a detailed description of the problem and any corrective action needed below.)		

Corrective Action–Use the space provided below to describe any problems identified during the inspection, along with a description of any corrective action needed. Appropriate action must be taken to ensure that the deficiencies are corrected. This may involve making changes to the procedure, providing retraining to employees, and/or taking other steps needed to ensure compliance.

Person(s) Conducting the Inspection		
Name (print):	**Signature:**	**Title/Department:**

Corrective Action

The lockout/tagout standard also requires that any deficiencies identified during the periodic inspection be corrected. Deficiencies, for example, may include deviation from established energy control procedures, inadequacies in an employee's knowledge, abilities, or use of the procedures, or inadequacies in the procedure itself. Following the periodic inspection, therefore, the inspector must report back to the employer with regard to any corrective action that is needed. The employer must then take appropriate action to ensure that the deficiencies are corrected. This may involve making changes to the procedure, providing retraining to employees, and/or taking other steps needed to ensure compliance.

Energized Testing or Positioning

Under certain circumstances, machines or equipment may need to be reenergized during service or maintenance. For example, testing with the power on is often necessary to ensure the proper assembly and functioning of components. Reenergization may also be required to perform troubleshooting or to position components. In these special situations, where the equipment has been locked/tagged out but reenergization is essential, the lockout/tagout standard allows for the *temporary* removal of locks or tags and the reenergization of the machine or equipment if a specific sequence of steps is followed. This sequence of steps is set forth in paragraph (f)(1) of the standard:

1910.147 (f)(1). *Testing or positioning of machines, equipment or components thereof.* In situations in which lockout or tagout devices must be temporarily removed from the energy isolating device and the machine or equipment energized to test or position the machine, equipment or component thereof, the following sequence of actions shall be followed:

i. Clear the machine or equipment of tools and materials in accordance with paragraph (e)(1) of this section;

ii. Remove employees from the machine or equipment area in accordance with paragraph (e)(2) of this section;

iii. Remove the lockout or tagout devices as specified in paragraph (e)(3) of this section;

iv. Energize and proceed with testing or positioning;

v. Deenergize all systems and reapply energy control measures in accordance with paragraph (d) of this section to continue the servicing and/or maintenance.

The prescribed sequence of actions is intended to ensure that lockout/tagout protection is maintained to the extent feasible during the energized testing or positioning. These provisions do not mean that lockout/tagout of the equipment during other portions of the servicing or maintenance task can be disregarded; locks/tags may be removed and the equipment reenergized only during the limited time necessary for the testing or positioning of the equipment or components. After the completion of the task for which reenergization was needed, the equipment must again be de-energized and the lockout/tagout procedures and protection resumed.

Performance of certain testing, troubleshooting, or positioning tasks may require that the equipment be alternately energized and de-energized, and that the sequence of steps be implemented several times. It is during such transition times when the servicing employees are alternately exposed to de-energized and energized conditions that the greatest risk of injury exists for them. Where energized testing or positioning under lockout/tagout is necessary, therefore, the written energy control procedure for the machine or equipment being serviced should reflect the required sequence of actions to be taken, and employees should be thoroughly trained in use of the procedure.

Emergency Lockout/Tagout Device Removal

The removal an employee's lock or tag from an energy isolating device without their knowledge could result in exposure to serious hazards that were assumed to be controlled. As a general rule, therefore, the lockout/tagout standard requires that a lockout or tagout device be removed only by the person who applied it. This basic requirement is found in paragraph (e)(3) of the standard:

> 1910.147 (e)(3) *Lockout or tagout devices removal.*
>
> Each lockout or tagout device shall be removed from each energy isolating device by the employee who applied the device.

This provision serves to assure the employee who is working under the protection of the lock and/or tag that the device will remain in place and not be removed by another person for the duration of the service or maintenance task. This control by authorized employees over their locks and/or tags and the corresponding energy isolating devices is one of the core principles of any lockout/tagout program.

In promulgating the standard, however, OSHA realized that special situations could arise where the employee who attached the device might not be available to remove it, and incorporated provisions that permit an employer to remove an employee's lockout or tagout device when certain specific conditions can be met. These provisions are set forth in a noted exception to paragraph (e)(3):

> Exception to paragraph (e)(3):
>
> When the authorized employee who applied the lockout or tagout device is not available to remove it, that device may be removed under the direction of the employer, provided that specific procedures and training for such removal have been developed, documented and incorporated into the employer's energy control program.

> The employer shall demonstrate that the specific procedure provides equivalent safety to the removal of the device by the authorized employee who applied it. The specific procedure shall include at least the following elements:
>
> (i) Verification by the employer that the authorized employee who applied the device is not at the facility:
>
> (ii) Making all reasonable efforts to contact the authorized employee to inform him/her that his/her lockout or tagout device has been removed; and
>
> (iii) Ensuring that the authorized employee has this knowledge before he/she resumes work at that facility.

Situations where an employee who applied a lockout or tagout device would not be available to remove it, for example, could result from the sudden sickness or injury of the employee, or other emergencies that result in the employee's absence from the workplace. Another scenario where device removal by the employer might be justified could occur when an employee who has locked/tagged out a piece of equipment is traveling or otherwise off site and it becomes *urgent* to return the equipment back to service, but it is impractical for the employee to return to the work site. Removal of an employee's lock and/or tag could also be necessary due to termination of the employee, abandonment of the device, or cases where the owner of the device is not identifiable.

Removal of a personal lockout or tagout device by another person must not, however, be based on convenience or simple unavailability of the employee; rather, it must be based on a demonstrated need to remove the device. Under no circumstances should an individual's lock or tag be removed by another employee when he/she is on site and available to remove the devices him/herself. Certainly, proper planning can help reduce or eliminate the need for removal of devices by persons other than those who applied them.

In addition, an employer may use the exception for removing an employee's personal lockout or tagout device only if the written energy control program includes specific procedures for such removal, and employees have been trained in the use of those procedures. The three steps outlined in paragraph (e)(3) represent reasonable and prudent measures for ensuring that the employee who applied the device for his or her protection is not exposed to energy hazards after its removal. The employer must be able to demonstrate that the removal procedure provides a level of safety equivalent to having the employee remove his/her own device.

The means by which an employer can actually remove lockout or tagout devices are limited, and depend on how the devices are managed. For example, where individually keyed locks are issued to authorized employees under the "one lock, one key" approach, it is necessary to use bolt cutters or similar tools to accomplish removal. Of course, this results in the destruction of the device. Alternatively, a master or duplicate key may be used by the employer to remove a lock, as long as the energy control program includes effective measures for controlling access to the master or duplicate keys. For tagout devices (where used alone), there is no equivalent master key concept. The attachment means for tags, pursuant to the standard, must be of a non-reusable, self-locking, and non-releasable type, and removal requires the destruction of the means of attachment (typically a one-piece, self-locking nylon cable tie).

OSHA's interpretive guidance on such use of the master keys is as follows:

> Prior to the use of the master key method, specific procedures and training, meeting the §1910.147(e)(3) exception, must be developed, documented, and incorporated into [the] energy control program. Among the features essential to a compliant master key procedure is a reliable method to ensure that access to the master key will be carefully controlled by the employer such that only those persons authorized and trained to use the master key in accordance with the employer's program can gain access. Safety is ensured not through the use of a specific removal device, be it a master key or bolt-cutter; rather, it lies in effective procedures, careful training, and procedures designed to ensure accountability. The success of any employer's energy control program, including lockout or tagout device removal actions, depends upon ensuring that its employees follow established, effective procedures, thereby respecting the sanctity of another employee's lockout or tagout device.

An example worksheet for special lockout/tagout device removal is provided below (see Figure 10.3). Example procedures for lock and/or tag removal by an employee other than the one who applied it are outlined in the example written lockout/tagout policy found in Appendix D.

Protective Materials and Hardware

To perform service or maintenance safely and effectively under lockout/tagout, employees must be provided with protective equipment, hardware, and appliances for isolating, securing, or blocking of machines or equipment from their energy sources. Under paragraph (c)(5) of the OSHA standard, it is the employer's responsibility to provide the necessary devices. Protective materials and hardware that may be needed include locks, tags, cable ties, chains, breaker and valve lockouts, wedges, blocks, blinds, pins, fasteners, and other hardware.

> 1910.147 (c)(5) *Protective materials and hardware.*
>
> (i) Locks, tags, chains, wedges, key blocks, adapter pins, self-locking fasteners, or other hardware shall be provided by the employer for isolating, securing or blocking of machines or equipment from energy sources.

With respect to providing the devices to employees, one option is to issue devices directly to each authorized employee who would personally use and maintain them. Another option is for the employer to simply maintain a sufficient quantity of the devices on hand and issue them to employees as the need arises. The protective materials and hardware must be available to authorized employees whenever they are needed to accomplish the service or maintenance tasks.

When launching a lockout/tagout program, a survey of the workplace is needed, not only to identify the machines, equipment, and processes that are covered by the program, but also to become familiar with the configuration of equipment, systems, and their energy isolating devices, so that an estimate of the types and quantities of protective hardware needed can be made. When ordering the necessary hardware, the employer must also ensure that the hardware meets specific requirements for design and construction. In general, lockout and tagout devices must be durable, standardized, substantial, and identifiable. The majority of needed hardware and appliances is available from original equipment manufacturers and third party vendors. In rare circumstances, devices may have to be custom fabricated.

Figure 10.3. Special Lock and Tag Removal Worksheet.

Special Lock and Tag Removal Worksheet	
Requester's Name:	Date:
Signature:	
I desire to remove the lock and/or tag presently securing the following equipment:	
The name and department of the tagger (person who applied the lock and tag) is:	
I request authority to remove the lock and tag because:	
I have considered the following safety items:	
Approval Signatures At least one signature required in each of the two groups below	
Requester's Supervisor:	Date:
Requester's Department Manager:	Date:
Tagger's Supervisor:	Date:
Tagger's Department Manager:	Date:
Owner or Primary Manager of Affected Equipment:	Date:

Post a copy of this worksheet on or closely adjacent to the equipment that was locked until the tagger has been informed, or until all reasonable efforts to identify the lock owner have been exhausted.

Source: US Dept. of Energy, Thomas Jefferson National Accelerator Facility, Newport New, VA.

Specific performance requirements for the design and construction of lockout and tagout devices are prescribed in paragraph (c)(5)(ii) and (c)(5)(iii) of the standard:

1910.147 (c)(5)(ii)

Lockout devices and tagout devices shall be singularly identified; shall be the only devices(s) used for controlling energy; shall not be used for other purposes; and shall meet the following requirements:

(A) Durable.

(1) Lockout and tagout devices shall be capable of withstanding the environment to which they are exposed for the maximum period of time that exposure is expected.

(2) Tagout devices shall be constructed and printed so that exposure to weather conditions or wet and damp locations will not cause the tag to deteriorate or the message on the tag to become illegible.

(3) Tags shall not deteriorate when used in corrosive environments such as areas where acid and alkali chemicals are handled and stored.

(B) Standardized. Lockout and tagout devices shall be standardized within the facility in at least one of the following criteria: Color; shape; or size; and additionally, in the case of tagout devices, print and format shall be standardized.

(C) Substantial

(1) Lockout devices. Lockout devices shall be substantial enough to prevent removal without the use of excessive force or unusual techniques, such as with the use of bolt cutters or other metal cutting tools.

(2) Tagout devices. Tagout devices, including their means of attachment, shall be substantial enough to prevent inadvertent or accidental removal. Tagout device attachment means shall be of a non-reusable type, attachable by hand, self-locking, and non-releasable with a minimum unlocking strength of no less than 50 pounds and having the general design and basic characteristics of being at least equivalent to a one-piece, all environment-tolerant nylon cable tie.

(D) Identifiable. Lockout devices and tagout devices shall indicate the identity of the employee applying the device(s).

1910.147 (c)(5)(iii)

Tagout devices shall warn against hazardous conditions if the machine or equipment is energized and shall include a legend such as the following: Do Not Start. Do Not Open. Do Not Close. Do Not Energize. Do Not Operate.

Paragraph (c)(5)(ii) of the standard specifies that lockout or tagout devices be "singularly identified," "the only devices used for controlling hazardous energy," and "not be used for other purposes." Using the devices for other purposes (such as using the designated locks on lockers or toolboxes), for example, could diminish the respect that employees must have for them. In addition, section (B) of (c)(5)(ii) requires that lockout or tagout devices be standardized in either color, shape, size, print, or format. The requirement for standardized devices is meant to ensure that they are readily identifiable, unique, and distinguished from other similar devices found in the workplace. The intent of these provisions is to elevate the importance of lockout/tagout devices and ensure that all employees understand their distinct and unmistakable purpose—to protect service and maintenance employees from accidental startup or the release of stored energy.

Sections (A) through (D) of paragraph (c)(5)(ii) outline specific performance requirements for lockout/tagout devices; that is, that they be durable, standardized, and substantial, and that they identify the user. The durability requirements are intended to ensure that the lockout and tagout devices will not fail to perform their intended job. Locks (and any adapters used for attaching them to the isolating device) must be substantial enough to prevent their removal by unauthorized persons without the use of excessive force or special techniques. Locks and adapters purchased from any reputable manufacturer generally provide adequate durability.

Tags (when used alone) and their means of attachment must be constructed to minimize the potential for them to be disconnected, lost, or accidentally removed. Tag attachment means are further required to be attachable by hand, and equivalent in strength to a one-piece non-releasable, self-locking cable tie. To meet these requirements, a standard ¼-inch nylon cable tie that has an unlocking strength of about 50 pounds is typically used for attaching tags. In addition, tags must be constructed and printed so that exposure to weather or other environmental conditions that may exist where they are placed will not cause them to become unserviceable or the message on the tag to become illegible. Using laminated tags with reinforcing grommets is generally adequate for the vast majority of work environments.

Section (c)(5)(ii)(D) of the standard requires that lockout or tagout devices identify the employee who applies them. This requirement adds a degree of accountability to the lockout/tagout program. For example, listing the identity of the user on lockout/tagout devices enables

Figure 10.4. Lock with attached tag (left); lock with collar and metal tag (right).

Figure 10.5. Examples of acceptable warnings for major message on tagout device. (Courtesy of Brady Signmark)

the employer and other employees to determine which authorized employees are performing a given servicing operation. Once the involved employee is identified, he/she can then be located if needed. It may be necessary, for example, to locate the identified employee to ensure that he/she is clear of danger before restarting the equipment, or to inquire about energy control procedures or practices during a periodic inspection. The identification also enables the employer to quickly determine the identity of an employee who may have abandoned or forgot to remove a lockout/tagout device. Identification can be accomplished by means of a collar and engraved tag attached to the lock, a durable label on the lock with the name of the employee, or by using a tag with the lock that identifies the employee (see Figure 10.4). As an added measure, color-coding of locks is often used to identify employees by group, trade, department, and so on.

In paragraph (c)(5)(iii), OSHA requires that the major message printed on the tag "warn against hazardous conditions if the equipment is re-energized." Five examples of acceptable warnings are provided, including: Do Not Start, Do Not Open, Do Not Close, Do Not Energize, and Do Not Operate (see Figure 10.5). Other warning messages, including those that use graphics, pictographs or other symbols are also acceptable if they convey the warning message in an equivalent manner. In all cases, employees must know and understand the meaning of the tag.

Machine/Equipment Design and Installation

One of the major provisions of the lockout/tagout standard is a requirement for machines and equipment to be retrofitted to accept locks (rendered capable of being locked out) if necessary, when any major replacement, repair, renovation, or modification occurs. This requirement is found in paragraph (c)(2)(iii) of the standard:

> 1910.147 (c)(2)(iii)
>
> After January 2, 1990, whenever replacement or major repair, renovation or modification of a machine or equipment is performed, and whenever new machines or equipment are installed, energy isolating devices for such machine or equipment shall be designed to accept a lockout device.

It is widely accepted that the use of locks is a more secure and more effective means for ensuring that equipment will not be reenergized while employees are working on it. The intent of the above provision, therefore, is to ensure that existing equipment which is not designed to be locked out, will have lockout capability at some time in the future if it remains in service. Once capable of being locked out, the equipment is then subject to the requirement to use the more protective system of lockout, except when a tagout system can be shown to be equally effective. When lockout capability is designed into new equipment, companies are more likely to employ a lockout system rather than dealing with the added restrictions of a tagout system.

To meet these requirements, employers may need to replace or modify components, or to purchase or fabricate appliances that enable the energy isolating devices to accept a lock. In some cases, it may be possible to buy or make an adapter that allows an isolating device (that lacks an integral means for attaching a lock) to be secured with a chain, block, brace, etc., that can then be locked. With some older valves and electrical disconnects that won't accept a lockout device, however, total replacement of the component may be the only viable option. Where the replacement or modification of equipment is necessary to achieve lockout capability, it is generally most effectively and efficiently accomplished as part of the normal replacement cycle of the equipment.

In addition, any new or replacement machines or equipment that is ordered or purchased (after January 2, 1990) must also be provided with the capability of being locked out, if such design is feasible. The vast majority of equipment manufacturers are now providing built-in lockout capability, or offer lockout devices specific to their equipment as accessories. When buying new equipment, managers and purchasing personnel should ensure not only that new equipment is "capable of being locked out," but also that ease of lockout is incorporated into the equipment's design when possible.

High Voltage Considerations

Although serious or fatal injuries can occur at low voltages (electrical *current* is really the culprit), the risk of serious injury or death grows as voltage increases. Voltages over 600 volts, for example, can quickly break down human skin. Once the skin is ruptured, the body's resistance is greatly reduced, allowing large current flow, which causes greater damage to tissues and internal organs. Equipment and circuits operated at higher than 600 volts, nominal, are generally considered to be high voltage. Most electric utilization equipment, including industrial machines, motors, pumps, fixtures, and appliances, operate at voltages lower than 600 volts.

Special equipment and systems, and equipment used in power generation, transmission, and distribution, however, often use voltages well above 600 volts.

When dealing with high voltages, employees may face a risk of electric shock and burns due to exposure to hazardous differences in electrical potential. These differences in potential can cause electrical arcing, which is caused by the flow of current through a path created by the vaporization of the terminal material. Arcs can occur as a result of faults (for example, created by poor electrical contact or failed insulation), lightning strikes, or inadvertent activation. The temperature of a high voltage electrical arc at the terminals can reach 35,000 degrees; to put this in perspective, 9,000 degrees is the temperature of the sun's surface. Consequently, a high voltage arc can kill up to 8 feet, and burn up to 20 feet away. About half of all serious electrical injuries involve burns from arcs. Additionally, *electric blasts* can occur as a result of the rapid expansion of air and gases caused by the electrical arc. The blast develops high pressures and is powerful enough to throw people and equipment, and damage or destroy structures. Personnel that work on or around high voltage circuits and equipment need to be made aware of these potential hazards and trained in the safety-related work practices needed to prevent and/or avoid them.

Figure 10.6. Specialized personal protective equipment and test instrumentation are needed for safely performing work on high voltage circuits and equipment.

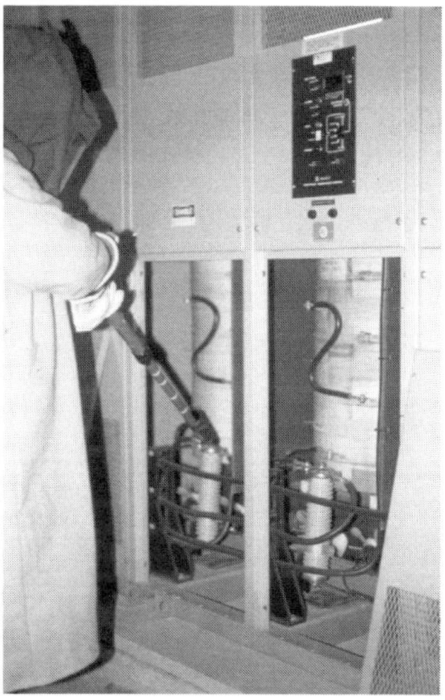

Because of the inherently greater hazards of high voltage, special procedures are needed for safely performing lockout/tagout-related activities such as energy isolation and control, verification, energized testing, and restoration of equipment to service. These may include: grounding; bonding; guarding; specialized test procedures and instruments; electrical protective clothing and equipment; specialized tools; and other measures (see Figure 10.6). These activities require special knowledge, education, and experience, and must only be performed by well-trained, qualified persons.

Switching Operations

A significant number of high voltage electrical accidents occur during switching operations (the closing or opening of electrical switchgear such as switches, circuit breakers, and contactors). Failure of these switching devices due to inadequate rating, overloading, or damage can cause faults that result in electric arcs or blasts. Hazards can also exist when restarting equipment after a fault. Resetting or replacing switches or breakers without correcting the underlying cause of the fault, for example, can result in equipment damage, and can increase the potential for arcs.

Before resetting or replacing the device, the reason that it tripped or operated should be determined. It is also important to shed loads from switching devices (to minimize current draw) where possible before reenergizing by ensuring that equipment on the load side of the circuit is turned off. Opening a switch not rated for load interruption while energized could also result in catastrophic failure of the device and lead to serious injury. To ensure personnel safety, switching orders are often developed and used to strictly regulate the operation of high voltage energy control devices (see Figure 10.7).

Figure 10.7. Caution sign reminds operators of precautions when performing switching.

Grounding

In some cases, there may be a potential for circuits or equipment that have been de-energized to be accidentally reenergized (due to faults, induced voltage, backfeed, and so on). As an additional protective measure, therefore, temporary grounding is often required and used in conjunction with lockout/tagout to protect employees who work on high voltage circuits and equipment from reenergization and arcing hazards (see Figure 10.8). The purpose of grounding is two-fold: to facilitate operation of the overcurrent device, and to equalize large voltage (potential) differences so that arcing does not occur. With proper grounding, if a fault occurs, a large amount of current can flow to ground, and the resulting surge will trip the overcurrent device (for example, a breaker), de-energizing the circuit.

Figure 10.8. Protective ground installed on a transformer.

To ensure that it is safe to install a protective ground, the circuit or equipment must first be tested and verified as dead. When a ground is to be attached to a line or to equipment, the ground-end connection must be attached first before attaching the other end. When a ground is to be removed, the grounding device must be removed from the line or equipment before the ground-end connection is removed. An insulated live-line tool (a "hot stick") must generally be used when installing or removing grounds. These tools insulate the employee from energized parts and enable the employee to work at a safe distance.

Testing, Inspection and Maintenance of Electrical Equipment

To ensure the safe and proper operation of electrical equipment, procedures for the testing, inspection, and preventive maintenance should be established and followed. Various special tests involving high voltage circuits and equipment may be needed, for example, as part of installation acceptance, routine inspection or maintenance, troubleshooting, or to otherwise ensure proper system operation (for example, after repairs). These include cable-fault locating, large capacitive load tests, high current fault-closure tests, insulation resistance and leakage tests, direct-current proof tests, phasing, and other tests (see Figure 10.9). Appropriate action must be taken to ensure that any problems identified during testing or inspection are corrected. In this regard, the equipment manufacturer should be consulted for appropriate guidance on testing, cleaning, recalibration, and repairs. The National Electrical Manufacturers Association also publishes various standards related to the handling, installation, operation, and maintenance of electrical equipment, and should also be consulted as necessary.

Electrical Standards

To protect employees from electrical hazards, OSHA prescribes electrical safety requirements in Subpart S, 29 CFR 1910.302 through 1910.399 of its General Industry Safety and Health Standards, including 1910.333, which deals with the *Selection and Use of Electrical Work Practices.* Subpart S contains requirements that apply to all electrical installations and utilization equipment, regardless of when they were designed or installed.

For construction operations, electrical installation safety requirements are prescribed in Subpart K, 29 CFR 1926.402 through 1910.408 of OSHA's Construction Safety and Health Standards. 29 CFR 1910.269 of Subpart R addresses the work practices to be used during the operation and maintenance of electric power generation, transmission, and distribution facilities. In addition to OSHA regulations, NFPA 70—the *National Electrical Code* (NEC®) and the *National Electrical Safety Code* (NESC) also address electrical safety requirements.

The NESC contains rules necessary for the protection of persons during the installation, operation, or maintenance of electric supply and communication lines and associated equipment. These rules contain basic provisions for protecting employees, and include work practices in addition to installation requirements.

Figure 10.9. Phasing is used to ensure that electric lines are properly installed in-phase after repairs.

The NEC® contains detailed provisions for the design and installation of electric conductors and equipment within or on buildings and other structures, and serves as the basis for electrical building codes across the United States.

A companion to the NEC® is NFPA 70E – *Electrical Safety Requirements for Employee Workplaces*. This document sets forth requirements for the practical safeguarding of employees from electrical hazards. Specific sections of NFPA 70E include: Installation Safety Requirements (Part I); Safety Related Work Practices (Part II); Safety Related Maintenance Requirements (Part III), and Safety Requirements for Special Equipment (Part IV).

Chapter 11

Beyond OSHA Compliance

The provisions of OSHA's *Control of Hazardous Energy* standard prescribe the *minimum* requirements that employers must follow for implementing energy control in the workplace. Many companies, however, recognize opportunities for achieving a higher level of safety, and supplement their lockout/tagout programs with additional elements that go beyond basic compliance. These additional elements involve effective alternative measures, equipment design, employee involvement, enhanced documentation and training, and human factors.

Maintenance personnel often face severe time constraints for returning equipment to service, a fact that adds to the difficulty of the repair work. A key for creating a sustainable program, therefore, is to make it as easy as possible for employees to implement. Building lockout capability into machines and having lockout materials and hardware readily available for use, for example, can save time and effort (see Figure 11.1). While not explicitly required by any rulemaking, the demonstrated support of management and the promotion of safe work habits and attitudes are elements that form the framework for any successful lockout/tagout program. Management can demonstrate their support, for example, by assigning responsibility, dedicating adequate resources, providing thorough, quality training, devising excellent energy control procedures, and by otherwise placing a strong emphasis on the lockout/tagout program.

Figure 11.1. Lockout stations can provide employees with convenient access to lockout/tagout materials and hardware. (Photo courtesy of Brady Signmark)

Machine Safeguarding and Effective Alternative Measures

Having to perform lockout on a machine each time a routine service or maintenance task such as lubrication, cleaning, unjamming, or adjusting is performed can be time consuming and can adversely affect production. If the task can be performed in a way that prevents employee exposure to energy hazards, however, such as by the use of special tools and/or alternative procedures that provide effective protection, it may not be necessary to apply the standard. If workers aren't exposed to hazards, and as long as guarding is not removed or bypassed, locking out of the equipment would not be required. Eliminating hazards with this approach can help companies maintain or improve productivity while still assuring the safety of personnel.

Figure 11.2. A "bottle prompter" is used to clear a jam between hydraulically-actuated molds in a blowmolding machine.

Consider the unjamming of machines such as conveyors, packagers, and labelers. In a production environment, jams can occur frequently, requiring repetitive clearing. Unjamming a machine is considered "service or maintenance" under the OSHA standard, and lockout is required each time an employee performs this function if employees are exposed to injury or must bypass safety guards. Since machine operating controls are generally not considered to be isolating devices, lockout is usually accomplished at the local disconnect or main breaker (assuming electrical energy is the only type of energy encountered). To do this each time the machine jams is burdensome and time consuming, and can slow production. In such cases, a disincentive to follow lockout/tagout may be established, and workers may feel pressure to take risks. Often, however, special tools, guarding, or other alternative measures can be provided to prevent workers from contacting hazardous machine parts during the unjamming operation (see Figure 11.2). If it can be demonstrated that the alternative means enables the servicing employee to clean, unjam, or otherwise service the machine without being exposed to hazards, the task is considered part of "normal production operations," and lockout is not required.

Similarly, the lubrication of power transmission apparatus and other moving parts is a frequent and necessary maintenance task. If the apparatus is guarded (for example, with a fixed enclosure guard), the lubrication task will likely require removal of the guard to gain access. If employees are exposed to injury as a result, the equipment must be de-energized and locked

Figure 11.3. Porthole in a fixed guard on calender roll power transmission apparatus allows remote oiling.

out. Providing a small porthole in a guard or enclosure can often allow for the remote lubrication of equipment, making guard removal unnecessary (see Figure 11.3). As long as the guarding adequately protects employees from injury, lockout is not required. Look for opportunities to engineer hazards out of machines; both safety and productivity can be increased.

Energy Control Procedures

Vital to program success is developing procedures that are understandable and easy for employees to implement. Putting together good procedures, however, often requires significant thought and effort. While it may be possible to use generic procedures and be in compliance, a lack of detail can make them tougher for employees to comprehend and follow. Detailed, equipment-specific procedures, on the other hand, can provide clear instructions for accomplishing energy control, eliminate the chance for mistakes, and often reduce the time it takes to accomplish the service or maintenance task.

Under the periodic inspection requirements of the standard, energy control procedures must be reviewed at least annually to ensure that they are adequate, updated, and accessible. If any deficiencies in the procedures are identified during the annual review, they must be corrected. Although yearly reviews may suffice for compliance purposes, it simply makes sense for deficiencies to be corrected whenever they are identified or there is reason to believe that they exist. If continuous improvement of the lockout/tagout program is sought, updating of procedures should be done not only when deficiencies exist, but also when improvements to the procedures are identified.

The written energy control procedures should be readily accessible to employees. For machines and equipment that require frequent service or maintenance, it's a good practice to post the energy control procedures at or near the machines to be serviced so that they are handy (see Figure 11.4). Clear plastic sheets will help protect the documents from damage. Computer work stations that host electronic versions of energy control procedures can also provide convenient access.

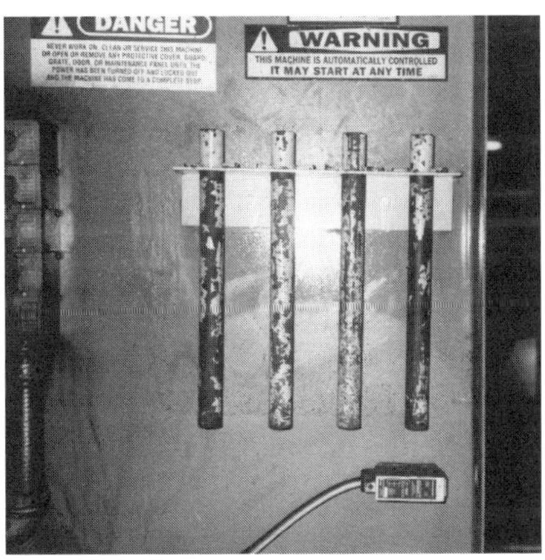

Figure 11.4. (Left) Posting the energy control procedure at the equipment for easy access is a best management practice.

Figure 11.5. (Right) Blocking pins are kept at a palletizer to ensure that they are readily accessible.

Similarly, protective materials and hardware should be readily available, by placing lockout stations close to machines and equipment requiring frequent service or maintenance under lockout, or by personally assigning locks, tags, and appliances to authorized employees as with other types of personal protective equipment. If possible, any special lockout devices such as blocks, pins, or wedges that may be needed should be kept with or near the machine so that they are on hand (see Figure 11.5). Employees shouldn't be tempted to take shortcuts because they have to walk a distance to obtain lockout devices or retrieve copies of energy control procedures.

Employee Participation

Although the OSHA standard does not require employee participation in the development of the lockout/tagout program or energy control procedures, the input of employees should be sought for several reasons: a) they are the ones most affected by the procedures, and b) they are often the ones most familiar with the equipment and therefore a great source of information on how best to accomplish energy isolation and control. The active involvement of employees can have a number of benefits, including enhanced understanding of energy control procedures, and the promotion of a sense of ownership for the lockout/tagout program on the part of employees. In contrast, the *failure* to involve employees in the decisions that affect them is listed by quality expert J. Edwards Deming as one of the major errors that can be made by management.

Opportunities for employee involvement exist in many areas, including the development and/or revision of procedures, performance of inspections and observations, and the design and presentation of training. In addition, mentoring and supervision by experienced employees is essential for transferring needed skills and knowledge to new employees. The diverse backgrounds and experience of employees can also be an invaluable source of new or better ideas; listening to employees can help companies tap into this resource. Communication with and

Figure 11.6. Hazard identification and warning signs and labels can serve as a constant reminder to follow safe work practices. (Courtesy of Brady Signmark)

Figure 11.7. Labeling of energy isolation points can also help personnel with lockout procedures. (Courtesy of Brady Signmark)

feedback from employees should be sought continuously to identify deficiencies in the lockout program, as well as opportunities for improvement.

Equipment and System Design

The proper design and configuration of machines, equipment, and systems is one of the most important considerations for accomplishing effective energy control. It is not only important that machines and equipment are capable of being locked out, but also that lockout can be accomplished easily and conveniently. For example, the strategic placement of isolating devices such as breakers, disconnects, and valves in a system can save valuable time and effort during the lockout process, and is especially important when lockout is required frequently. Other measures such as the labeling of isolating devices and piping systems is also of great benefit to employees in assuring that energy controls can be quickly and properly applied (see Figures 11.6 and 11.7). Coordination between engineering and purchasing personnel is needed to ensure that proper safeguarding and lockout capability is built in to new (or extensively modified) machines and equipment. In addition, it is important to educate engineering and purchasing employees *even if not required under the standard* so that they have an understanding of what is required.

Electrical Equipment

With electrical utilization equipment such as motors, motor circuits, and controllers, OSHA regulations and the National Electrical Code® require that disconnecting means such as safety switches, local disconnects, and/or circuit breakers be provided to cut off the equipment from its source of supply. Generally speaking, the disconnecting means must be in sight from the machinery, motor, and/or controller, where "in sight from" means visible and not more than 50 feet from the specified equipment. In the case of motor branch circuits over 600 volts, nominal, the controller disconnecting means is permitted to be out of sight of the controller, if the controller is marked with a warning label giving the location and identification of the disconnecting means, which is to be locked in the open position.

Figure 11.8. Lockable disconnect switch provided on compactor machine.

While locating the disconnecting means within 50 feet may suffice for compliance purposes, that location may not necessarily be convenient for purposes of locking and tagging out the equipment, especially if service or maintenance under lockout is required frequently. For example, on a conveyor line, the disconnect may be on the side of the conveyor opposite from the operator work station or the location where a jam occurs. Even though the disconnect is in sight, the operator (authorized employee) may need to clamber over, under, or around the conveyor if a lockout/tagout situation is presented. In such cases, it may take more time to access the disconnect and secure it than it does to clear the actual jam. This is inconvenient, and again, presents a disincentive to the employee to properly de-energize and lock the equipment out.

Jams often occur in the same location on machines, equipment, or systems. When this is the case, it can be beneficial to place a disconnect close to the jam location. Convenient access to the disconnect makes it easier for employees to isolate the equipment, and thus they are more likely to use it (see Figure 11.8). Assuming that there are no other hazards, if the disconnect can be maintained under the exclusive control of the employee, it may also be possible to use the standard's minor servicing exception, so that locking/tagging of the device would not be required. "Under the exclusive control of the employee" means that the disconnect is located such that the employee would be able to prevent it from being operated (closed) during the servicing or maintenance task (typically, within arm's reach).

Piping Systems

The locking/tagging out of piping systems used for the distribution of compressed air, hydraulic fluid, steam, water, and chemicals can sometimes be challenging, especially where isolating means are limited. With piping systems, therefore, it is important that a sufficient number of control and/or sectioning valves be provided so that the piping segments or components of concern can be locked/tagged out without difficulty. To ensure that any residual energy in the piping system can be removed or relieved, appropriate bleeds, vents, drains, and flares should be installed in strategic locations. Similarly, appropriate gauges and other devices for indicating pressure, flow, level, temperature, and so on should placed at key points in the piping system so that the authorized employee can easily verify that de-energization and isolation have been accomplished.

When dealing with highly hazardous chemicals, the use of automatic valves that can be controlled remotely should be considered to minimize the potential exposure of employees. If pipe openings are expected, the locations where service, maintenance, or repairs are likely to be needed should be anticipated, and design features (for example, barriers, deflectors, and containment) incorporated, if possible, to minimize the potential exposure of personnel to spills or releases that result from disengagement or breaking of lines.

Labeling of Energy Isolating Devices

In promulgating the lockout/tagout standard, OSHA initially proposed incorporating provisions that would require the marking or labeling of energy isolating devices used for the control of potentially hazardous energy sources, unless the devices are positioned and arranged so that their purpose is evident. The proposed identification requirement would have applied to all energy isolating devices, including valves or similar devices used to control hydraulic, pneumatic, steam, chemical, and similar energy sources. In the final rule, however, OSHA eliminated this proposed requirement, reasoning that other provisions (those requiring authorized employees to have specific knowledge about the type, magnitude, and hazards of the energy to be encountered, and the method or means to control the energy *before* servicing the machine or equipment) were sufficient for ensuring personnel safety.

For electrical equipment, a similar requirement is largely in place already. For example, OSHA's Subpart S regulations and the National Electrical Code® require that each disconnecting means for motors and appliances be correctly and legibly marked to indicate its purpose, unless located and arranged so the purpose is evident. In addition, each service, feeder, and branch circuit must be correctly and legibly marked at its disconnecting means or overcurrent device to indicate its purpose. Although not *required* for other isolating devices such as valves, labeling can make it easier for employees to identify the correct devices needed for energy isolation, and verify that energy controls have been properly applied. As a good management practice, therefore, labels, tags, or other forms of marking should be provided on energy isolating devices that serve as key control points for blocking the flow of energy (see Figures 11.9 and 11.10). Labels should identify the equipment supplied and the substance conveyed. Often, it is useful to assign a unique identification number to the valve and include this information on the marking or label.

Figure 11.9. Electrical disconnect labeled to indicate its purpose.

Figure 11.10. Identification tag used on emergency shut-off valve for hydrochloric acid in a chemical injection system.

Similarly, the marking or labeling of piping systems, although not required by the standard, is an invaluable aid to workers for properly identifying piping system contents when pipe openings or other covered tasks are performed. Identification may be accomplished, for example, by stenciling, the use of tape, or markers. ASME A13.1-1996 provides a *Scheme for the Identification of Piping Systems*, and should be followed if possible. Pursuant to ASME A13.1, legends should be applied close to valves or flanges and adjacent to changes in direction, branches, and where pipes pass through walls or floors, and at intervals on straight pipe runs sufficient for identification. The legend should identify the contents of the piping system, indicate direction of flow, and provide sufficient additional details, such as temperature and pressure, necessary to identify the hazard.

Preventative Maintenance

Ideally, the exposure of personnel to hazardous energy is prevented by eliminating or minimizing the need for service or maintenance. In general, the proper initial design, construction, installation, and operation of equipment and systems can reduce the potential for premature failure and the need for major repairs.

Nevertheless, normal wear and tear on machines, equipment, and components will occur, requiring service or maintenance at some point. The occurrence or severity of breakdowns, and the frequency of needed service or repairs, however, can often be reduced with a program of regular preventative maintenance. Often, keeping machines lubricated and properly adjusted is all that is needed to maintain smooth and reliable operation. Preferably, machine design will allow routine lubrication and adjustment without removal of safeguards.

Equipment operators are typically most familiar with the functioning and performance of the equipment that they work with, and can often detect early warning signs of potential trouble. Operators, for example, may notice unusual noises, vibrations, leaks, deterioration, excessive operating temperatures or pressures, or other signs that might signal a potential problem. Personnel should be alert for these early warning signs, and report any problems observed or suspected to their supervisor or maintenance department so that appropriate action can be taken. Prompt attention to minor problems may prevent a more serious problem (and thus more involved service or maintenance) from developing. If an unsafe condition exists, the equipment should be taken out of service immediately until repairs are made.

Mechanical Integrity Program

For processing operations that involve the manufacture, use, storage, or handling of highly hazardous chemicals above certain threshold quantities, OSHA's Process Safety Management regulations (29 CFR 1910.119) require the establishment of a Mechanical Integrity Program. The purpose of the program is to assure that equipment used to handle highly hazardous chemicals is designed, constructed, installed, and maintained to minimize the risk of chemical spills or releases. Such equipment includes:

- pressure vessels and storage tanks;
- piping systems and components such as valves;
- relief and vent systems and devices;
- emergency shutdown systems;
- controls (including monitoring devices and sensors, alarms and interlocks); and
- pumps.

Under the program, equipment inspections, testing and maintenance must be performed to identify problems and ensure that deficiencies are corrected. Required elements of a Mechanical Integrity Program include:

- the identification and categorization of equipment and instrumentation;

- development of written maintenance procedures;
- training for process maintenance activities;
- inspection and testing;
- correction of deficiencies in equipment that are outside acceptable limits defined by the process safety information; and
- development of a quality assurance program.

Although the requirement for establishing a Mechanical Integrity Program applies only to operations involving highly hazardous chemicals, many other operations can benefit from similar programs as well.

Chapter 12

Epilogue

With the pressure and stress of modern life, it's easy for workers to get careless. When working with industrial machinery, a moment of carelessness can easily result in serious injury or death. The greatest benefit of a high performance lockout/tagout program is that it provides a system of checks and balances to guard against carelessness and protect employees against injury. Implementing lockout is not merely a matter of compliance; it is also a demonstration of management's care and concern for the safety and welfare of employees, and a good business practice. Companies that fail to recognize and control machine hazards face regulatory fines, and risk the occurrence of a serious and costly machine-related accident. World-class companies maintain effective lockout programs to prevent these predictable losses.

Appendix A

29 CFR 1910.147—The Control of Hazardous Encrgy (Lockout/Tagout)

(a) Scope, application and purpose.

(1) Scope.

(i) This standard covers the servicing and maintenance of machines and equipment in which the unexpected energization or start up of the machines or equipment, or release of stored energy could cause injury to employees. This standard establishes minimum performance requirements for the control of such hazardous energy.

(ii) This standard does not cover the following:

(A) Construction, agriculture and maritime employment;

(B) Installations under the exclusive control of electric utilities for the purpose of power generation, transmission and distribution, including related equipment for communication or metering; and

(C) Exposure to electrical hazards from work on, near, or with conductors or equipment in electric utilization installations, which is covered by Subpart S of this part; and

(D) Oil and gas well drilling and servicing.

(2) Application.

(i) This standard applies to the control of energy during servicing and/or maintenance of machines and equipment.

(ii) Normal production operations are not covered by this standard (See Subpart O of this Part). Servicing and/or maintenance which takes place during normal production operations is covered by this standard only if:

(A) An employee is required to remove or bypass a guard or other safety device; or

(B) An employee is required to place any part of his or her body into an area on a machine or piece of equipment where work is actually performed upon the material being processed (point of operation) or where an associated danger zone exists during a machine operating cycle

Exception to paragraph (a)(2)(ii): Minor tool changes and adjustments, and other minor servicing activities, which take place during normal production operations, are not covered by this standard if they are routine, repetitive, and integral to the use of the equipment for production, provided that the work is performed using alternative measures which provide effective protection (See Subpart O of this Part).

(iii) This standard does not apply to the following:

(A) Work on cord and plug connected electric equipment for which exposure to the hazards of unexpected energization or start up of the equipment is controlled by the unplugging of the equipment from the energy source and by the plug being under the exclusive control of the employee performing the servicing or maintenance.

(B) Hot tap operations involving transmission and distribution systems for substances such as gas, steam, water or petroleum products when they are performed on pressurized pipelines, provided that the employer demonstrates that:

(1) continuity of service is essential;

(2) shutdown of the system is impractical; and

(3) documented procedures are followed, and special equipment is used which will provide proven effective protection for employees.

(3) Purpose.

(i) This section requires employers to establish a program and utilize procedures for affixing appropriate lockout devices or tagout devices to energy isolating devices, and to otherwise disable machines or equipment to prevent unexpected energization, start up or release of stored energy in order to prevent injury to employees.

(ii) When other standards in this part require the use of lockout or tagout, they shall be used and supplemented by the procedural and training requirements of this section.

(b) Definitions applicable to this section.

Affected employee. An employee whose job requires him/her to operate or use a machine or equipment on which servicing or maintenance is being performed under lockout or tagout, or whose job requires him/her to work in an area in which such servicing or maintenance is being performed.

Authorized employee. A person who locks out or tags out machines or equipment in order to perform servicing or maintenance on that machine or equipment. An affected employee becomes an authorized employee when that employee's duties include performing servicing or maintenance covered under this section.

Capable of being locked out. An energy isolating device is capable of being locked out if it has a hasp or other means of attachment to which, or through which, a lock can be affixed, or it has a locking mechanism built into it. Other energy isolating devices are capable of being locked out, if lockout can be achieved without the need to dismantle, rebuild, or replace the energy isolating device or permanently alter its energy control capability.

Energized. Connected to an energy source or containing residual or stored energy.

Energy isolating device. A mechanical device that physically prevents the transmission or release of energy, including but not limited to the following: A manually operated electrical circuit breaker; a disconnect switch; a manually operated switch by which the conductors of a circuit can be disconnected from all ungrounded supply conductors, and, in addition, no pole can be operated independently; a line valve; a block; and any similar device used to block or isolate energy. Push buttons, selector switches and other control circuit type devices are not energy isolating devices.

Energy source. Any source of electrical, mechanical, hydraulic, pneumatic, chemical, thermal, or other energy.

Hot tap. A procedure used in the repair, maintenance and services activities which involves welding on a piece of equipment (pipelines, vessels or tanks) under pressure, in order to install connections or appurtenances. it is commonly used to replace or add sections of pipeline without the interruption of service for air, gas, water, steam, and petrochemical distribution systems.

Lockout. The placement of a lockout device on an energy isolating device, in accordance with an established procedure, ensuring that the energy isolating device and the equipment being controlled cannot be operated until the lockout device is removed.

Lockout device. A device that utilizes a positive means such as a lock, either key or combination type, to hold an energy isolating device in the safe position and prevent the energizing of a machine or equipment. Included are blank flanges and bolted slip blinds.

Normal production operations. The utilization of a machine or equipment to perform its intended production function.

Servicing and/or maintenance. Workplace activities such as constructing, installing, setting up, adjusting, inspecting, modifying, and maintaining and/or servicing machines or equipment. These activities include lubrication, cleaning or unjamming of machines or equipment and making adjustments or tool changes, where the employee may be exposed to the unexpected energization or startup of the equipment or release of hazardous energy.

Setting up. Any work performed to prepare a machine or equipment to perform its normal production operation.

Tagout. The placement of a tagout device on an energy isolating device, in accordance with an established procedure, to indicate that the energy isolating device and the equipment being controlled may not be operated until the tagout device is removed.

Tagout device. A prominent warning device, such as a tag and a means of attachment, which can be securely fastened to an energy isolating device in accordance with an established procedure, to indicate that the energy isolating device and the equipment being controlled may not be operated until the tagout device is removed.

(c) General.

(1) Energy control program. The employer shall establish a program consisting of energy control procedures, employee training and periodic inspections to ensure that before any employee performs any servicing or maintenance on a machine or equipment where the unexpected energizing, startup or release of stored energy could occur and cause injury, the machine or equipment shall be isolated from the energy source and rendered inoperative.

(2) Lockout/tagout.

(i) If an energy isolating device is not capable of being locked out, the employer's energy control program under paragraph (c)(1) of this section shall utilize a tagout system.

(ii) If an energy isolating device is capable of being locked out, the employer's energy control program under paragraph (c)(1) of this section shall utilize lockout, unless the employer can demonstrate that the utilization of a tagout system will provide full employee protection as set forth in paragraph (c)(3) of this section.

(iii) After January 2, 1990, whenever replacement or major repair, renovation or modification of a machine or equipment is performed, and whenever new machines or equipment are installed, energy isolating devices for such machine or equipment shall be designed to accept a lockout device.

(3) Full employee protection.

(i) When a tagout device is used on an energy isolating device which is capable of being locked out, the tagout device shall be attached at the same location that the lockout device would have been attached, and the employer shall demonstrate that the tagout program will provide a level of safety equivalent to that obtained by using a lockout program.

(ii) In demonstrating that a level of safety is achieved in the tagout program which is equivalent to the level of safety obtained by using a lockout program, the employer shall demonstrate full compliance with all tagout-related provisions of this standard together with such additional elements as are necessary to provide the equivalent safety available from the use of a lockout device. Additional means to be considered as part of the demonstration of full employee protection shall include the implementation of additional safety measures such as the removal of an isolating circuit element, blocking of a controlling switch, opening of an extra disconnecting device, or the removal of a valve handle to reduce the likelihood of inadvertent energization.

(4) Energy control procedure.

(i) Procedures shall be developed, documented and utilized for the control of potentially hazardous energy when employees are engaged in the activities covered by this section.

Exception: The employer need not document the required procedure for a particular machine or equipment, when all of the following elements exist: (1) The machine or equipment has no potential for stored or residual energy or reaccumulation of stored energy after shut down which could endanger employees; (2) the machine or equipment has a single energy source which can be readily identified and isolated; (3) the isolation and locking out of that energy source will completely deenergize and deactivate the machine or equipment; (4) the machine or equipment is isolated from that energy source and locked out during servicing or maintenance; (5) a single lockout device will achieve a locked-out condition; (6) the lockout device is under the exclusive control of the authorized employee performing the servicing or maintenance; (7) the servicing or maintenance does not create hazards for other employees; and (8) the employer, in utilizing this exception, has had no accidents involving the unexpected activation or reenergization of the machine or equipment during servicing or maintenance.

(ii) The procedures shall clearly and specifically outline the scope, purpose, authorization, rules, and techniques to be utilized for the control of hazardous energy, and the means to enforce compliance including, but not limited to, the following:

(A) A specific statement of the intended use of the procedure;

(B) Specific procedural steps for shutting down, isolating, blocking and securing machines or equipment to control hazardous energy;

(C) Specific procedural steps for the placement, removal and transfer of lockout devices or tagout devices and the responsibility for them; and

(D) Specific requirements for testing a machine or equipment to determine and verify the effectiveness of lockout devices, tagout devices, and other energy control measures.

(5) Protective materials and hardware.

(i) Locks, tags, chains, wedges, key blocks, adapter pins, self-locking fasteners, or other hardware shall be provided by the employer for isolating, securing or blocking of machines or equipment from energy sources.

(ii) Lockout devices and tagout devices shall be singularly identified; shall be the only devices(s) used for controlling energy; shall not be used for other purposes; and shall meet the following requirements:

(A) Durable.

(1) Lockout and tagout devices shall be capable of withstanding the environment to which they are exposed for the maximum period of time that exposure is expected.

(2) Tagout devices shall be constructed and printed so that exposure to weather conditions or wet and damp locations will not cause the tag to deteriorate or the message on the tag to become illegible.

(3) Tags shall not deteriorate when used in corrosive environments such as areas where acid and alkali chemicals are handled and stored.

(B) Standardized. Lockout and tagout devices shall be standardized within the facility in at least one of the following criteria: Color; shape; or size; and additionally, in the case of tagout devices, print and format shall be standardized.

(C) Substantial:

(1) Lockout devices. Lockout devices shall be substantial enough to prevent removal without the use of excessive force or unusual techniques, such as with the use of bolt cutters or other metal cutting tools.

(2) Tagout devices. Tagout devices, including their means of attachment, shall be substantial enough to prevent inadvertent or accidental removal. Tagout device attachment means shall be of a non-reusable type, attachable by hand, self-locking, and non-releasable with a minimum unlocking strength of no less than 50 pounds and having the general design and basic characteristics of being at least equivalent to a one-piece, all environment-tolerant nylon cable tie.

(D) Identifiable. Lockout devices and tagout devices shall indicate the identity of the employee applying the device(s).

(iii) Tagout devices shall warn against hazardous conditions if the machine or equipment is energized and shall include a legend such as the following: Do Not Start. Do Not Open. Do Not Close. Do Not Energize. Do Not Operate.

(6) Periodic inspection.

(i) The employer shall conduct a periodic inspection of the energy control procedure at least annually to ensure that the procedure and the requirements of this standard are being followed.

(A) The periodic inspection shall be performed by an authorized employee other than the ones(s) utilizing the energy control procedure being inspected.

(B) The periodic inspection shall be conducted to correct any deviations or inadequacies identified.

(C) Where lockout is used for energy control, the periodic inspection shall include a review, between the inspector and each authorized employee, of that employee's responsibilities under the energy control procedure being inspected.

(D) Where tagout is used for energy control, the periodic inspection shall include a review, between the inspector and each authorized and affected employee, of that employee's responsibilities under the energy control procedure being inspected, and the elements set forth in paragraph (c)(7)(ii) of this section.

(ii) The employer shall certify that the periodic inspections have been performed. The certification shall identify the machine or equipment on which the energy control procedure was being utilized, the date of the inspection, the employees included in the inspection, and the person performing the inspection.

(7) Training and communication.

(i) The employer shall provide training to ensure that the purpose and function of the energy control program are understood by employees and that the knowledge and skills required for the safe application, usage, and removal of the energy controls are acquired by employees. The training shall include the following:

(A) Each authorized employee shall receive training in the recognition of applicable hazardous energy sources, the type and magnitude of the energy available in the workplace, and the methods and means necessary for energy isolation and control.

(B) Each affected employee shall be instructed in the purpose and use of the energy control procedure.

(C) All other employees whose work operations are or may be in an area where energy control procedures may be utilized, shall be instructed about the procedure, and about the prohibition relating to attempts to restart or reenergize machines or equipment which are locked out or tagged out.

(ii) When tagout systems are used, employees shall also be trained in the following limitations of tags:

(A) Tags are essentially warning devices affixed to energy isolating devices, and do not provide the physical restraint on those devices that is provided by a lock.

(B) When a tag is attached to an energy isolating means, it is not to be removed without authorization of the authorized person responsible for it, and it is never to be bypassed, ignored, or otherwise defeated.

(C) Tags must be legible and understandable by all authorized employees, affected employees, and all other employees whose work operations are or may be in the area, in order to be effective.

(D) Tags and their means of attachment must be made of materials which will withstand the environmental conditions encountered in the workplace.

(E) Tags may evoke a false sense of security, and their meaning needs to be understood as part of the overall energy control program.

(F) Tags must be securely attached to energy isolating devices so that they cannot be inadvertently or accidentally detached during use.

(iii) Employee retraining.

(A) Retraining shall be provided for all authorized and affected employees whenever there is a change in their job assignments, a change in machines, equipment or processes that present a new hazard, or when there is a change in the energy control procedures.

(B) Additional retraining shall also be conducted whenever a periodic inspection under paragraph (c)(6) of this section reveals, or whenever the employer has reason to believe that there are deviations from or inadequacies in the employee's knowledge or use of the energy control procedures.

(C) The retraining shall reestablish employee proficiency and introduce new or revised control methods and procedures, as necessary.

(iv) The employer shall certify that employee training has been accomplished and is being kept up to date. The certification shall contain each employee's name and dates of training.

(8) Energy isolation. Lockout or tagout shall be performed only by the authorized employees who are performing the servicing or maintenance.

(9) Notification of employees. Affected employees shall be notified by the employer or authorized employee of the application and removal of lockout devices or tagout devices. Notification shall be given before the controls are applied, and after they are removed from the machine or equipment.

(d) Application of control. The established procedures for the application of energy control (the lockout or tagout procedures) shall cover the following elements and actions and shall be done in the following sequence:

(1) Preparation for shutdown. Before an authorized or affected employee turns off a machine or equipment, the authorized employee shall have knowledge of the type and magnitude of the energy, the hazards of the energy to be controlled, and the method or means to control the energy.

(2) Machine or equipment shutdown. The machine or equipment shall be turned off or shut down using the procedures established for the machine or equipment. An orderly shutdown must be utilized to avoid any additional or increased hazard(s) to employees as a result of the equipment stoppage.

(3) Machine or equipment isolation. All energy isolating devices that are needed to control the energy to the machine or equipment shall be physically located and operated in such a manner as to isolate the machine or equipment from the energy source(s).

(4) Lockout or tagout device application.

(i) Lockout or tagout devices shall be affixed to each energy isolating device by authorized employees.

(ii) Lockout devices, where used, shall be affixed in a manner to that will hold the energy isolating devices in a "safe" or "off" position.

(iii) Tagout devices, where used, shall be affixed in such a manner as will clearly indicate that the operation or movement of energy isolating devices from the "safe" or "off" position is prohibited.

(A) Where tagout devices are used with energy isolating devices designed with the capability of being locked, the tag attachment shall be fastened at the same point at which the lock would have been attached.

(B) Where a tag cannot be affixed directly to the energy isolating device, the tag shall be located as close as safely possible to the device, in a position that will be immediately obvious to anyone attempting to operate the device.

(5) Stored energy.

(i) Following the application of lockout or tagout devices to energy isolating devices, all potentially hazardous stored or residual energy shall be relieved, disconnected, restrained, and otherwise rendered safe.

(ii) If there is a possibility of reaccumulation of stored energy to a hazardous level, verification of isolation shall be continued until the servicing or maintenance is completed, or until the possibility of such accumulation no longer exists.

(6) Verification of isolation. Prior to starting work on machines or equipment that have been locked out or tagged out, the authorized employee shall verify that isolation and deenergization of the machine or equipment have been accomplished.

(e) Release from lockout or tagout. Before lockout or tagout devices are removed and energy is restored to the machine or equipment, procedures shall be followed and actions taken by the authorized employee(s) to ensure the following:

(1) The machine or equipment. The work area shall be inspected to ensure that nonessential items have been removed and to ensure that machine or equipment components are operationally intact.

(2) Employees.

(i) The work area shall be checked to ensure that all employees have been safely positioned or removed.

(ii) After lockout or tagout devices have been removed and before a machine or equipment is started, affected employees shall be notified that the lockout or tagout device(s) have been removed.

(3) Lockout or tagout devices removal. Each lockout or tagout device shall be removed from each energy isolating device by the employee who applied the device. Exception to paragraph (e)(3): When the authorized employee who applied the lockout or tagout device is not available to remove it, that device may be removed under the direction of the employer, provided that specific procedures and training for such removal have been developed, documented and incorporated into the employer's energy control program. The employer shall demonstrate that the specific procedure provides equivalent safety to the removal of the device by the authorized employee who applied it. The specific procedure shall include at least the following elements:

(i) Verification by the employer that the authorized employee who applied the device is not at the facility:

(ii) Making all reasonable efforts to contact the authorized employee to inform him/her that his/her lockout or tagout device has been removed; and

(iii) Ensuring that the authorized employee has this knowledge before he/she resumes work at that facility.

(f) Additional requirements.

(1) Testing or positioning of machines, equipment or components thereof. In situations in which lockout or tagout devices must be temporarily removed from the energy isolating device and the machine or equipment energized to test or position the machine, equipment or component thereof, the following sequence of actions shall be followed:

(i) Clear the machine or equipment of tools and materials in accordance with paragraph (e)(1) of this section;

(ii) Remove employees from the machine or equipment area in accordance with paragraph (e)(2) of this section;

(iii) Remove the lockout or tagout devices as specified in paragraph (e)(3) of this section;

(iv) Energize and proceed with testing or positioning;

(v) Deenergize all systems and reapply energy control measures in accordance with paragraph (d) of this section to continue the servicing and/or maintenance.

(2) Outside personnel (contractors, etc.).

(i) Whenever outside servicing personnel are to be engaged in activities covered by the scope and application of this standard, the on-site employer and the outside employer shall inform each other of their respective lockout or tagout procedures.

(ii) The on-site employer shall ensure that his/her employees understand and comply with the restrictions and prohibitions of the outside employer's energy control program.

(3) Group lockout or tagout.

(i) When servicing and/or maintenance is performed by a crew, craft, department or other group, they shall utilize a procedure which affords the employees a level of protection equivalent to that provided by the implementation of a personal lockout or tagout device.

(ii) Group lockout or tagout devices shall be used in accordance with the procedures required by paragraph (c)(4) of this section including, but not necessarily limited to, the following specific requirements:

(A) Primary responsibility is vested in an authorized employee for a set number of employees working under the protection of a group lockout or tagout device (such as an operations lock);

(B) Provision for the authorized employee to ascertain the exposure status of individual group members with regard to the lockout or tagout of the machine or equipment and

(C) When more than one crew, craft, department, etc. is involved, assignment of overall job-associated lockout or tagout control responsibility to an authorized employee designated to coordinate affected work forces and ensure continuity of protection; and

(D) Each authorized employee shall affix a personal lockout or tagout device to the group lockout device, group lockbox, or comparable mechanism when he or she begins work, and shall remove those devices when he or she stops working on the machine or equipment being serviced or maintained.

(4) Shift or personnel changes. Specific procedures shall be utilized during shift or personnel changes to ensure the continuity of lockout or tagout protection, including provision for the orderly transfer of lockout or tagout device protection between off-going and oncoming employees, to minimize exposure to hazards from the unexpected energization or start-up of the machine or equipment, or the release of stored energy.

The following appendix to §1910.147 services as a non-mandatory guideline to assist employers and employees in complying with the requirements of this section, as well as to provide other helpful information. Nothing in the appendix adds to or detracts from any of the requirements of this section.

Appendix A to 29 CFR 1910.147

General

The following simple lockout procedure is provided to assist employers in developing their procedures so they meet the requirements of this standard. When the energy isolating devices are not lockable, tagout may be used, provided the employer complies with the provisions of the standard which require additional training and more rigorous periodic inspections. When tagout is used and the energy isolating devices are lockable, the employer must provide full employee protection (see paragraph (c)(3)) and additional training and more rigorous periodic inspections are required. For more complex systems, more comprehensive procedures may need to be developed, documented, and utilized.

Lockout Procedure

Lockout Procedure for

(Name of Company for single procedure or identification of equipment if multiple procedures are used).

Purpose

This procedure establishes the minimum requirements for the lockout of energy isolating devices whenever maintenance or servicing is done on machines or equipment. It shall be used to ensure that the machine or equipment is stopped, isolated from all potentially hazardous energy sources and locked out before employees perform any servicing or maintenance where the unexpected energization or start-up of the machine or equipment or release of stored energy could cause injury.

Compliance With This Program

All employees are required to comply with the restrictions and limitations imposed upon them during the use of lockout. The authorized employees are required to perform the lockout in accordance with this procedure. All employees, upon observing a machine or piece of equipment which is locked out to perform servicing or maintenance, shall not attempt to start, energize, or use that machine or equipment.

(Type of compliance enforcement to be taken for violation of the above.)

Sequence of Lockout

1. Notify all affected employees that servicing or maintenance is required on a machine or equipment and that the machine or equipment must be shut down and locked out to perform the servicing or maintenance.

[Name(s)/Job Title(s) of affected employees and how to notify.]

2. The authorized employee shall refer to the company procedure to identify the type and magnitude of the energy that the machine or equipment utilizes, shall understand the hazards of the energy, and shall know the methods to control the energy.

[Type(s) and magnitude(s) of energy, its hazards and the methods to control the energy.]

3. If the machine or equipment is operating, shut it down by the normal stopping procedure (depress the stop button, open switch, close valve, etc.).

[Type(s) and location(s) of machine or equipment operating controls.]

4. De-activate the energy isolating device(s) so that the machine or equipment is isolated from the energy source(s).

[Type(s) and location(s) of energy isolating devices.]

5. Lock out the energy isolating device(s) with assigned individual lock(s).

6. Stored or residual energy (such as that in capacitors, springs, elevated machine members, rotating flywheels, hydraulic systems, and air, gas, steam, or water pressure, etc.) must be dissipated or restrained by methods such as grounding, repositioning, blocking, bleeding down, etc.

[Type(s) of stored energy—methods to dissipate or restrain.]

7. Ensure that the equipment is disconnected from the energy source(s) by first checking that no personnel are exposed, then verify the isolation of the equipment by operating the push button or other normal operating control(s) or by testing to make certain the equipment will not operate.

   Caution: Return operating control(s) to neutral or "off" position after verifying the isolation of the equipment.

[Method of verifying the isolation of the equipment.]

8. The machine or equipment is now locked out.

Restoring Equipment to Service.
When the servicing or maintenance is completed and the machine or equipment is ready to return to normal operating condition, the following steps shall be taken.

1. Check the machine or equipment and the immediate area around the machine to ensure that nonessential items have been removed and that the machine or equipment components are operationally intact.

2. Check the work area to ensure that all employees have been safely positioned or removed from the area.

3. Verify that the controls are in neutral.

4. Remove the lockout devices and reenergize the machine or equipment. Note: The removal of some forms of blocking may require reenergization of the machine before safe removal.

5. Notify affected employees that the servicing or maintenance is completed and the machine or equipment is ready for used.

[54 FR 36687, Sept. 1, 1989 as amended at 54 FR 42498, Oct. 17, 1989; 55 FR 38685, Sept. 20, 1990; 61 FR 5507, Feb. 13, 1996]

Appendix B

Lockout/Tagout Checklists

Lockout/Tagout Program Checklist

Criterion	Y	N	Comments
Has a workplace survey been performed to identify the machines, equipment, and processes that may expose workers to hazardous energy?			
Has a written lockout/tagout program been completed and is it readily available for review?			
Are means in place for enforcing the use of lockout/tagout procedures (e.g. disciplinary procedures)?			
Are equipment-specific energy control procedures completed before employees engage in covered service or maintenance tasks?			
Are the equipment-specific energy control procedures adequate?			
Are needed protective hardware and appliances readily available to employees?			
Have authorized, affected, and other employees been trained?			
When tagout alone is used, is a level of safety equivalent to that obtained by using lockout (i.e. "full employee protection") assured?			
Where tagout alone is used, are employees trained on the limitations of tags?			

Lockout/Tagout Program Checklist (Continued)

Criterion	Y	N	Comments
Are employees retrained whenever there is a change in their job assignments, a change in machines, equipment, or processes that present a new hazard, or a change in the energy control procedures?			
Are employees retrained whenever deviations from or inadequacies in an employee's knowledge or use of the energy control procedures are discovered?			
Is certification kept to document that lockout/tagout training has been accomplished?			
Are de-energization and isolation of machines and equipment verified prior to commencing covered service or maintenance tasks?			
Are affected employees notified of the application and removal of lockout and/or tagout devices?			
Are the proper procedures and sequence of actions followed during energized testing or positioning to ensure that lockout/tagout protection is maintained to the extent feasible?			
When group lockout or tagout is used, are procedures used that ensure a level of protection equivalent to that of individual lockout or tagout?			
Are specific procedures in place to ensure the continuity and orderly transfer of lockout/tagout protections during shift and/or personnel changes?			
Are lockout/tagout procedures coordinated with contractors and outside service personnel?			
Where duplicate or master keys are permitted for removal of lockout devices during emergencies, is access to the duplicate or master keys carefully controlled to prevent unauthorized access?			

Lockout/Tagout Program Checklist (Continued)

Criterion	Y	N	Comments
Where duplicate or master keys are permitted, have specific procedures and training for their use been developed, documented, and incorporated into the lockout/tagout program?			
Are periodic inspections of the energy control procedures conducted at least annually?			
Where lockout is used for energy control, does the periodic inspection include a review, between the inspector and each authorized employee, of that employee's responsibilities under the energy control procedure being inspected?			
Where tagout is used for energy control, does the periodic inspection include a review, between the inspector and each authorized and affected employee, of that employee's responsibilities under the energy control procedure being inspected?			
Are certifications completed to document that the periodic inspections have been performed?			
Are all deviations or inadequacies identified during the periodic inspections corrected?			
Are machines or equipment that are not capable of being locked out redesigned/retrofitted to accept a lock whenever major replacement, repair, renovation, or modification is performed?			

OSHA Lockout/Tagout Procedures Checklist

Procedure	
Is all machinery or equipment capable of movement, required to be de-energized or disengaged and locked out during cleaning, servicing, adjusting, or setting up operations, whenever required?	
Where the power disconnecting means for equipment does not also disconnect the electrical control circuit:	
Are the appropriate electrical enclosures identified?	
Is means provided to assure the control circuit can also be disconnected and locked out?	
Is the locking out of control circuits in lieu of locking out main power disconnects prohibited?	
Are all equipment control valve handles provided with a means for locking out?	
Does the lockout procedure require that stored energy (mechanical, hydraulics, air, etc.) be released or blocked before equipment is locked out for repairs?	
Are appropriate employees provided with individually keyed personal safety locks?	
Are employees required to keep personal control of their keys while they have safety locks in use?	
Is it required that only the employee exposed to the hazard, place or remove the safety lock?	
Is it required that employees check the safety of the lockout by attempting a startup after making sure no one is exposed?	
Are employees instructed to always push the control circuit stop button immediately after checking the safety of the lockout?	
Is there a means provided to identify any or all employees who are working on locked out equipment by their locks or accompanying tags?	
Are a sufficient number of accident preventive signs or tags and safety padlocks provided for any reasonably foreseeable repair emergency?	
When machine operations, configuration or size requires the operator to leave his or her control station to install tools or perform other operations, and that part of the machine could move if accidentally activated, is such element required to be separately locked or blocked out?	
In the event that equipment or lines cannot be shut down, locked out and tagged, is a safe job procedure established and rigidly followed?	

Source: OSHA Small Business Outreach Training Program, May, 1997.

Appendix C

Frequently Asked Questions

(Source: OSHA)

Scope and Application

Who does this standard apply to?
General Industry workers performing servicing and maintenance on machines and equipment and who are exposed to the unexpected energization, startup, or release of hazardous energy.

What activities or operations are covered?

- Any source of mechanical, hydraulic, pneumatic, chemical, thermal, or other energy.
- Constructing, installing, setting up, adjusting, inspecting, modifying, maintaining and/or servicing machines or equipment, including lubrication, cleaning or unjamming of machines or equipment, and making adjustments or tool changes, where employees could be exposed to the unexpected energization or startup of the equipment or release of hazardous energy.
- Servicing and maintenance activities performed during normal production operations **if:**
  - An employee is required to remove or bypass machine guards or other safety devices, or
  - An employee is required to place any part of his or her body into a point of operation or into an area on a machine or piece of equipment where work is performed, or into the danger zone associated with the machine's operation.

Who does this standard not apply to?
General Industry workers performing servicing and maintenance on machines or equipment who are NOT exposed to the unexpected energization or startup of the machines or equipment, or the release of hazardous energy.

What activities and operations are not covered?

- Servicing and maintenance of equipment performed during normal production operations **if:**
  - the safeguarding provisions of Subpart O, and other applicable general industry standards are effective in preventing worker exposure to hazards created by the unexpected energization or startup of machines or equipment, or the release of energy.

- Minor tool changes and adjustments, and other minor servicing activities that take place during normal production operations which are routine, repetitive, and integral to the use of that production equipment, as long as workers are effectively protected by alternative measures which provide effective machine safeguarding protection (See Subpart O).
- Construction, agriculture, and maritime workers.
- Installations under the exclusive control of electric utilities for power generation, transmission, and distribution.
- Exposure to electrical hazards from work on, near, or with conductors or equipment in electric utilization installations.
- Oil and gas well drilling and servicing.
- Work on cord and plug connected electrical equipment, **if:**
  - The equipment is unplugged from the energy source and the authorized employee has exclusive control of the plug.
- Hot tap operations that involve transmission and distribution systems for gas, steam, water, or petroleum products on pressurized pipelines, **if:**
  - Continuity of service is essential, shutdown of the system is impractical, documented procedures are followed, and employees are effectively protected by special equipment.

Purpose

What is the purpose of the standard?
To prevent the unexpected energization or startup of machines and equipment, or release of stored energy, in order to prevent workplace injuries during service and maintenance operations.

How is this accomplished?
Employers must establish an energy control program, consisting of energy control procedures, employee training, and periodic inspections to ensure that before service and maintenance is performed, machines and equipment that could unexpectedly start up, become energized, or release stored energy, are isolated from their energy source(s) and rendered safe.

Definitions

What definitions are useful to understand how the lockout tagout standard will be applied?
- **Authorized employee:** An employee who locks or tags machines or equipment in order to perform servicing or maintenance.
- **Affected employee:** An employee who is required to use machines or equipment on which servicing is performed under the Lockout/Tagout standard or who performs other job responsibilities in an area where such servicing is performed.

- **Other employees:** All employees who are or may be in an area where energy control procedures may be utilized.
- **Capable of being locked out:** An energy-isolating device is considered capable of being locked out if it:
  - Is designed with a hasp or other means of attachment to which a lock can be affixed.
  - Has a locking mechanism built into it.
  - Can be locked without dismantling, rebuilding, or replacing the energy-isolating device or permanently altering its energy control capability.

- **Energized:** Machines and equipment are energized when they are connected to an energy source or they contain residual or stored energy.
- **Energy-isolating device:** A mechanical device that physically prevents the transmission or release or energy, including but not limited to the following: a manually operated electrical circuit breaker; a disconnect switch; a manually operated switch by which the conductors of a circuit can be disconnected from all ungrounded supply conductors and, in addition, no pole can be operated independently; a line valve; a block; and any similar device used to block or isolate energy. Push buttons, selector switches and other control circuit type devices are not energy isolating devices.
- **Energy source:** Any source of electrical, mechanical, hydraulic, pneumatic, chemical, thermal, or other energy.
- **Lockout:** The placement of a lockout device on an energy-isolating device, in accordance with an established procedure, ensuring that the energy-isolating device and the equipment being controlled cannot be operated until the lockout device is removed.
- **Lockout device**: Any device that uses positive means, such as a lock, blank flanges and bolted slip blinds, to hold an energy-isolating device in a safe position, thereby preventing the energizing of machinery or equipment.
- **Normal production operations:** Utilization of a machine or equipment to perform its intended production function.
- **Servicing and/or maintenance:** Workplace activities such as constructing, installing, setting up, adjusting, inspecting, modifying, maintaining and/or servicing machines or equipment, including lubrication, cleaning or unjamming of machines or equipment, and making adjustments or tool changes, where employees could be exposed to the unexpected energization or startup of the equipment or release of hazardous energy.
- **Tagout:** The placement of a tagout device on an energy-isolating device, in accordance with an established procedure, to indicate that the energy-isolating device and the equipment being controlled may not be operated until the tagout device is removed.
- **Tagout device:** Any prominent warning device, such as a tag and a means of attachment, that can be securely fastened to an energy-isolating device to indicate that the machine or equipment to which it is attached may not be operated until the tagout device is removed.

Energy Control Program

What are the core components of the energy control program?

- Energy control procedures that detail and document the specific information that an authorized employee must know to accomplish lockout/tagout, namely, the scope, purpose, authorization rules and techniques to be utilized for the control of hazardous energy.
- Periodic inspections of the energy control procedures to ensure that the procedures and the requirements of the standard are being followed.
- Employee training and retraining, along with additional training under a tagout system, to ensure that the purpose and function of the energy control programs are understood by the employer.

What is the intent of the energy control program?

To ensure that before any employee services equipment where the potential exists for unexpected energization or start-up of equipment or the release of stored energy, the machine or equipment is isolated from the energy source and rendered inoperative.

Does the employer have the flexibility to develop his/her own program?

Yes. Employers are expected to develop programs and procedures, training and inspections, that meet the needs of their particular workplace and the particular types of machines and equipment they use and service as long as they meet the requirements of the standard.

Energy Control Procedures—Documentation

What is the employer's obligation in establishing energy control procedures?

Employers must develop, document, and use specific procedures to control potentially hazardous energy when employees are servicing equipment or machinery.

Under what limited situations is documentation of the procedures not required?

- The machine or equipment has no potential for stored or residual energy, or for reaccumulation of stored energy after shut down, which could endanger employees.
- The machine or equipment has a single energy source that can be readily identified and isolated and the isolation and locking out of that energy source will completely deenergize and deactivate the machine or equipment.
- The machine or equipment is isolated from that energy source and locked out during servicing or maintenance.
- A single lockout device will achieve a locked-out condition.
- The lockout device is under the exclusive control of the authorized employee performing the servicing or maintenance.
- The servicing or maintenance does not create hazards for other employees.

- The employer has had no incidents involving the unexpected activation or reenergization of machines or equipment during servicing or maintenance

Energy Control Procedures–Required Content

What specific elements must be documented in the employer's energy control procedures?

- The procedures must outline the scope, purpose, authorization, rules and techniques that the employer will use to control hazardous energy.
- The procedures must state the means to be used to enforce compliance.
- At a minimum, the procedures must include:
  - A specific statement of the intended use of the procedure.
  - Specific procedural steps for shutting down, isolating, blocking, and securing machines or equipment to control hazardous energy.
  - Specific procedural steps for the placement, removal, and transfer of lockout devices or tagout devices, and a description of who has responsibility for them.
  - Specific requirements for testing a machine or piece of equipment to determine and verify the effectiveness of lockout devices, tagout devices, and other energy control measures.

Periodic Inspection

What is the intent of the requirement for the employer to conduct periodic inspections?
To ensure that the energy control procedures continue to be implemented properly, that the employees are familiar with their responsibilities, and that any deviations or procedural inadequacies that are observed are corrected.

How often must the inspection take place?
At least annually.

Who performs the periodic inspection?
An authorized employee not involved in the energy control procedure being inspected.

What does the periodic inspection entail?

- The employer must identify any deficiencies or deviations and correct them.
- Where lockout is used, the inspector must review each authorized employee's responsibilities under the procedure with that employee (group meetings are acceptable).
- Where tagout is used, the inspector must review *both* the authorized and affected employee's responsibilities with those employees for the energy control procedure being inspected, and the additional training responsibilities of 1910.147(c)(7)(ii).
- The employer must certify that the periodic inspections have been performed.

What must the certification identify?

- The machine on which the procedure was utilized.

- The date of inspection.
- The identity of the employees included in inspection.
- The identity of the person who performed the inspection.

Employee Training And Communication

Why must employees affected by this standard be trained?

- So that they understand the purpose and function of the energy control program.
- So that employees acquire the knowledge and skills necessary for the safe application, usage and removal of the energy controls.

The standard requires different levels of training for the three categories of employees; what are the differences in the training required for the three categories?

- Authorized employees must receive training on the recognition of applicable hazardous energy sources, the type and magnitude of the energy available in the workplace, and the methods and means necessary for energy isolation and control.
- Affected employees must receive training on the purpose and use of the energy control procedure.
- Other employees (those whose work activities are or may be in an area where energy control procedures may be utilized) must be instructed about the procedure and about the prohibition relating to attempts to restart or reenergize machines or equipment that are locked out or tagged out.

Additional Training

What additional training is required when tagout systems are used?

- Employers must train employees in the following limitations of tags:
  - Tags are essentially warning devices affixed to energy isolating devices and do not provide the physical restraint on those devices that is provided by a lock.
  - When a tag is attached to an energy isolating means, it is not to be removed without authorization and it is never to be bypassed, ignored, or otherwise defeated.
  - Tags must be legible and understandable by all employees.
  - Tags and their means of attachment must be made of materials which will withstand the environmental conditions encountered in the workplace.
  - Tags may evoke a false sense of security and their meaning needs to be understood as part of the overall energy control program.
  - Tags must be securely attached to energy isolating devices so that they cannot be inadvertently or accidentally detached during use.

Employee Retraining

Who must be retrained?
All affected and authorized employees must be retrained under certain conditions listed below.

Is training required annually?
No.

What triggers the retraining requirements?
- A change in job assignments.
- A change in machines, equipment, or processes that present a new hazard.
- A change in the energy control procedures.
- Periodic inspections reveal that there are deviations in the energy control procedure.
- The employer believes that there are deviations from, or inadequacies in, the employee's knowledge or use of the energy control procedures.

What is the object of the retraining?
- To introduce new or revised control methods and procedures as necessary.
- To reestablish employee proficiency.

Does training require certification?
Yes. Employer must certify that training or retraining took place and that the employee is kept up to date.

What information must appear on the certificate?
- Each employee's name.
- The dates of training and/or retraining.

Lockout/Tagout

If an energy isolating device is not capable of being locked out, can the employer use a tagout system?
Yes, when the energy isolating devices are not lockable, tagout may be used, provided the employer complies with the provisions of the standard which require additional training and more rigorous periodic inspections.

If an energy isolating device is capable of being locked out, must the employer use a lock out system?
Yes. Unless the employer can show that the tagout system provides full employee protection, as described in paragraph (c)(3) of the standard.

New or Modified Equipment

What is the date after which all new machines and equipment, or all machines and equipment that undergo major repair, renovations or modification, must be equipped with energy-isolating devices capable of accepting a lockout device?
January 2, 1990.

Full Employee Protection

What are the requirements for the use of tagout devices when lockout devices are capable of being used?
- The tags are attached where the lockout devices would be.
- The employer demonstrates that the tagout will provide protection at least as effective as locks and will assure full employee protection.

How does an employer demonstrate that the protection achieved using the tagout program is equivalent to the level of safety obtained by using a lockout program?
The employer must comply with all tagout-related provisions and also use additional safety measures that provide a level of safety equivalent to that obtained by using lockout. This might include removing and isolating a circuit element, blocking a controlling switch, opening an extra disconnecting device, or removing a valve handle to reduce the potential for any inadvertent energization while the tags are attached.

Materials and Hardware

What protective materials and hardware must be provided by the employer for isolating, securing or blocking machines or equipment from energy sources?
Locks, tags, chains, wedges, key blocks, adapter pins, self-locking fasteners, or other hardware.

What are the requirements for the lockout and tagout devices?
- Must be durable, so that they are capable of withstanding the environment to which they are exposed for the maximum period of time that exposure is expected.
- Must be singularly identified.
- Must be the only devices used for controlling energy.
- Must not be used for other purposes.
- Must be standardized withing the facility in at least one of the following criteria: color, shape, or size. Additionally, tagout devices must be standardized as to print and format.
- Must be identifiable, in that it indicates the identity of the employee applying the devices.

In addition to the above, what other hardware requirements are specific to lockout?
Must be substantial enough to prevent removal without the use of excessive force or unusual techniques such as with the use of bolt cutters or other metal cutting tools.

In addition to the above, what other hardware requirements are specific to tagout?
- Must be constructed and printed so that exposure to weather conditions or wet and damp locations will not cause the tag to deteriorate or the message on the tag to become illegible.
- Must not deteriorate when used in corrosive environments such as areas where acid and alkali chemicals are handled and stored.
- Must be standardized in print and format.
- Must be substantial to prevent inadvertent or accidental removal.
- Must have an attachment means of a non-reusable type, attachable by hand, self-locking, and non-releasable with a minimum unlocking strength of no less than 50 pounds and having the general design and basic characteristics of being at least equivalent to a one-piece all-environment-tolerant nylon cable tie.
- Must warn against hazardous conditions if the machine or equipment is energized.
- Must include a legend such as: Do Not Start, Do Not Open, Do Not Close, Do Not Energize, Do Not Operate.

Application of Energy Control

To safely apply energy control to machines or equipment (using either lockout or tagout devices), authorized employees must perform certain procedures, in a specific order. What are the sequential procedures?
- **Preparation for shutdown:** Before an authorized or affected employee turns off a machine or equipment, the authorized employee must have knowledge of the type and magnitude of the energy, the hazards of the energy to be controlled, and the method or means to control the energy.
- **Machine or equipment shutdown:** The machine or equipment must be turned off or shut down using the procedures established for it to avoid any additional or increased hazards to employees as a result of the machine or equipment stoppage.
- **Machine or equipment isolation:** All energy-isolating devices that are needed to control the machine's energy source must be located. These devices must then be used to isolate the machine or equipment from its energy source.
- **Lockout or tagout device application:** Lockout or tagout devices must be affixed to each energy-isolating device by authorized employees. Lockout devices where used, must be affixed in a manner that will hold the energy isolating devices in a "safe" or "off" position. Where tagout devices are used, it must be affixed in a manner that will clearly indicate that the operation or movement of energy isolating devices from the "safe" or "off" position is prohibited. If the tag can not be affixed directly to the energy isolating device, the tag must be

located as close as safely possible to the device, in a position that will be immediately obvious to anyone attempting to operate the device.

- **Stored energy:** After the energy-isolating device has been locked out or tagged out, all potentially hazardous stored or residual energy must be relieved, disconnected, restrained, or otherwise rendered safe.

- **Verification of isolation:** Before any work begins on machines or equipment that have been locked out or tagged out, an authorized employee must verify that the machine or equipment has been properly isolated and deenergized.

Release From Lockout/Tagout

The Lockout/Tagout standard includes requirements for releasing machines or equipment that have been locked out or tagged out prior to restoring energy to the equipment and using it. Before lockout or tagout devices are removed, and energy restored, what procedures must the authorized employee follow?

- **Machine/equipment inspection:** The work area must be inspected to ensure that nonessential items (e.g., tools, spare parts) have been removed and that all of the machine or equipment components are operationally intact.

- **Positioning of employees:** The work area must be checked to ensure that all employees have been safely positioned or have cleared the area. In addition, all affected employees must be notified that the lockout or tagout devices have been removed before the equipment is started.

- **Lockout or tagout device removal:** Each lockout or tagout device must be removed from the energy-isolating device by the employee who applied the device.

What is the unique circumstance that allows an employee other than the one who applied the lockout/tagout device to remove the device?
When the authorized employee who applied the lockout or tagout device is not available to remove it, that device may be removed under the direction of the employer, provided that specific procedures and training for such removal have been developed, documented, and incorporated into the employer's energy control program.

What steps must the employer take if an employee, other than the one who applied the lockout/tagout device, removes the device?

- The employer must verify that the authorized employee who applied the device is not at the facility.

- The employer must make all reasonable efforts to contact the authorized employee to inform him/her that his/her lockout or tagout device has been removed.

- The employer must ensure that the authorized employees knows that the lockout device has been removed before he/she resumes work at the facility.

Testing of Machines

When may lockout or tagout devices be removed temporarily?
In some circumstances, employees need to temporarily restore energy to a machine or piece of equipment during servicing or maintenance to test and /or reposition the machine or piece of equipment. Lockout or tagout devices may be removed temporarily in order to perform these tasks.

What sequence of action must occur in the temporary removal of the lockout/tagout devices?

1. The machine or equipment must be cleared of tools and materials.
2. Employees must be removed from the machine or equipment area.
3. All lockout or tagout devices may then be removed.
4. Authorized employees may then proceed to energize and test or position the equipment or machinery.
5. Following testing or positioning, all systems must be deenergized and energy control measures reapplied to continue the servicing and /or maintenance.

Outside Personnel (Contractors)

What are the obligations of the outside contractor and the on-site employer?
- Whenever contractors and other outside servicing personnel perform tasks covered by the Lockout/Tagout standard, they must adhere to all the standard's requirements.
- The contractor or outside employer and the on-site employer must inform each other of their respective energy control program responsibilities.
- The on-site employer must ensure that his/her employees understand and comply with the restrictions and prohibitions of the outside employer's energy control program.

Group Lockout/Tagout Requirements

Can servicing or maintenance be performed by a crew, department, or other group under this standard?
Yes. If they have been properly trained and the energy control program is followed.

What procedures must be followed that will offer group employees the same protection that the standard provides to individual employees?
- A group lockout/tagout must afford each employee a level of protection equivalent to that provided by the implementation of a personal lockout or tagout device.
- Primary responsibility for a set number of employees working under the protection of a group lockout or tagout device must be vested in a single authorized employee.

- The single authorized employee must determine the exposure status of individual group members.
- If there will be more than one crew, department, or group involved in the activity, a single authorized employee must be designated to coordinate affected workforces and to ensure continuity of protection.
- Each authorized employee must affix a personal lockout or tagout device to the machine or equipment when work begins and remove it when work is completed.

Shift and Personnel Changes

How is the continuity of lockout or tagout protection maintained during shift or personnel changes?
Employers must ensure the continuity of employee protection by providing for the orderly transfer of lockout or tagout device protection between off-going and incoming employees. This will help to minimize exposure to hazards from the unexpected energization or start-up of the machine or equipment or the release of stored energy.

Appendix D

Sample Lockout/Tagout Program

1.0. PURPOSE

ABC Company maintains a comprehensive occupational safety and health program and is committed to the protection of employees who are potentially exposed to machine, equipment and process (MEP) hazards. Under this program, hazardous energy associated with machines, equipment and processes will be controlled during covered service and maintenance activities. This program also establishes the procedures for ABC Company's Energy Control Program in compliance with OSHA regulations for the control of hazardous energy (lockout/tagout). The failure of employees to comply with the provisions of this policy may result in disciplinary action up to and including discharge.

2.0. APPLICATION

1. This procedure applies to all temporary, contract, and permanent employees, and visitors who may be engaged in the following:
   - (a) servicing or maintaining equipment in which the unexpected energization or start-up of a machine, equipment or process, or release of stored energy could cause injury to employees;
   - (b) operating equipment that is being serviced or maintained, and
   - (c) working in an area in which a lockout or tagout system is being implemented.
2. This program does not apply to "normal production operations" when alternative measures which provide effective employee protection are used, unless:
   - (a) an employee is required to remove or bypass a guard or other safety device, or
   - (b) an employee is required to place any part of his/her body into an area of the machine or equipment where he/she could be injured.
3. This program does not apply to work on cord and plug connected equipment where:
   - (a) the unexpected energization or start-up is controlled by unplugging such equipment, and
   - (b) the plug is under the exclusive control of the employee performing the servicing or maintenance.
4. Other Relevant Policies. Although the control of electrical energy is generally addressed in this policy/procedure, EHS Policy/Procedure No. 009—Electrical Safety-Related Work Practices should also be consulted as necessary when dealing with electrical circuits and equipment.

3.0. DEFINITIONS

Affected employee.

An employee whose job requires him/her to operate or use a machine or equipment on which servicing or maintenance is being performed under lockout or tagout, or whose job requires him/her to work in an area in which such servicing or maintenance is being performed.

Authorized employee.

A person who locks out or tags out machines or equipment to perform the servicing or maintenance on that machine. An authorized employee and an affected employee may be the same person when the affected employee's duties also include performing maintenance or service on equipment that will be locked out or tagged out.

Capable of being locked out.

An energy isolating device will be considered to be "capable of being locked out" either if it is designed with a hasp or other means of attachment to which, or through which, a lock can be affixed, or if it has a locking mechanism built into it. Other energy isolating devices will also be considered to be capable of being locked out if lockout can be achieved without the need to dismantle, rebuild, or replace the energy isolating device or permanently alter its energy control capability.

Energy control procedure.

Detailed, equipment-specific procedures that supplement and work in conjunction with the general provisions of the overall written program, and describe: the machine, equipment or process to be locked/tagged out; the intended use of the procedure and/or type of service/maintenance to be performed; the types and magnitude of hazardous energy associated with the machine equipment or process (MEP); the type and location of the isolating device(s); involved personnel; the specific procedures and sequences for deenergizing, verifying, and locking/tagging out the MEP; and the specific steps for reenergizing the machine and restoring it back to normal operation. isolate energy.

Energy isolating device.

A mechanical device that physically prevents the transmission or release of energy, including but not limited to the following: a manually operated electrical circuit breaker; a disconnect switch; a manually operated switch by which the conductors of a circuit can be disconnected from all ungrounded supply conductors and, in addition, no pole can be operated independently; a slide gate; a line valve; a block; and any similar device used to block or isolate energy.

The term *does not include* a push button, selector switch, and other control circuit type devices.

Energy source.

Any source of electrical, mechanical, hydraulic, pneumatic, chemical, thermal or other energy.

Lockout

The placement of a lockout device on an energy isolating device in accordance with an established procedure, ensuring that the isolating device and the equipment being controlled cannot be operated until the lockout device is removed.

Lockout device.

A device that utilizes a positive means such as a lock, either key or combination type, to hold an energy isolating device in the safe position and prevent the energizing of a machine or equipment.

Normal Production operations.

The utilization of a machine or equipment to perform its intended production function.

Qualified person.

One familiar with the construction and operation of the equipment and the hazards involved.

Servicing and/or maintenance.

Workplace activities such as constructing, installing, setting up, adjusting, inspecting, modifying, and maintaining and/or servicing machines or equipment. These activities include lubrication, cleaning or unjamming of machines or equipment and making adjustments or tool changes, where the employee may be exposed to the unexpected energization or start-up of the equipment or release of hazardous energy.

Tagout

The placement of a tagout device on an energy isolating device, in accordance with established procedure, to indicate that the energy isolating device and equipment being controlled may not be operated until the tagout device is removed.

Tagout device.

A prominent warning device, such as a tag and a means of attachment, which can be securely fastened to an energy isolating device in accordance with an established procedure, to indicate that the energy isolating device and the equipment being controlled may not be operated until the tagout device is removed.

Zero Energy State.

That state of a machine, equipment or process in which all sources of hazardous energy are brought to their lowest practical value and locked/tagged out.

4.0. RESPONSIBILITIES

1. All personnel, including employees and contractor employees will comply with the provisions of this procedure.
2. Managers and Supervisors are responsible for:
   - (a) Surveying their work areas to identify the machines, equipment and processes (MEP) that utilize hazardous energy and that are covered by this procedure.
   - (b) Ensuring that equipment-specific energy control procedures are established for the machines, equipment and processes in their area.

(c) Ensuring that employees and contractors follow established lockout/tagout and energy control procedures.

(d) Ensuring that employees receive proper lockout/tagout training prior to assignment to covered service or maintenance tasks.

(e) Ensuring that only authorized employees perform service or maintenance under lockout/tagout.

(f) Ensuring that only qualified persons perform electrical-related work.

(g) Providing protective equipment, hardware and appliances (e.g., locks, lockout devices, keys, tags, etc.) conforming to the provisions of this procedure to employees requiring this equipment.

(h) Designating the authorized employee who has primary responsibility for overall lockout/tagout control in a group lockout situation.

(i) When necessary (i.e., in a group lockout situation when a "log and lock" approach is used), keeping a log of lockout/tagout device application and removal, including: the person applying lock/tag; purpose of lockout/tagout; anticipated duration of lockout/tagout; and actual date of removal of locks/tags.

(j) Maintaining exclusive control of duplicate or master keys.

(k) Ensuring that the periodic (e.g., annual) inspections are conducted and documented as required by section 8.0 of this procedure.

3. Authorized Employees are responsible for:

(a) Following the procedures that have been developed for lockout/tagout and ensuring the proper use of his/her equipment;

(b) Determining the type(s), magnitude(s) and hazards of energy to be controlled, and the method(s) or means to control the energy before commencing the covered service or maintenance task;

(c) Locating and identifying all energy isolating devices that will be locked or tagged out before commencing the covered service or maintenance task;

(d) Reporting any lost lockout/tagout equipment immediately to his/her Supervisor.

(e) Inspecting his/her own equipment prior to use;

(f) Notifying affected employees before lockout/tagout is performed and then again before energy is restored once servicing or maintenance is complete;

(g) Participating in the development of equipment-specific energy control procedures, and reporting any problems associated with those procedures; and

(h) Attending assigned training sessions.

4. Affected Employees will:

(a) Be aware of lockout/tagout procedures used to guard against unexpected startup.

(b) Not attempt to operate any MEP which is locked out or tagged out.

(c) Not attempt to remove any locks or tags from energy isolating devices.

5. Corporate EHS is responsible for:

(a) Developing and maintaining the program required by this procedure and the applicable standards.

(b) Providing guidance on the interpretation of this procedure.

(c) Assisting with provision of training to ensure compliance with this procedure.

(d) Auditing compliance with established ABC Company procedures.

5.0. PROCEDURES

5.1. General

1. During the servicing or maintenance of machines, equipment and processes in which unexpected energization, start up, or release of stored energy could occur and cause injury, all sources of hazardous energy will be deenergized, isolated or otherwise effectively controlled and locked/tagged out.
2. No one will attempt to operate any energy isolating device when it is locked or tagged out.
3. Lockout (vs. tagout) should be used when the energy isolating device is capable of being locked out. Tags will be affixed to the isolating devices along with locks, by placing the shank of the lock through the grommet provided on the tag.
4. Tagout, without lockout, will only be used when the energy isolating device is not capable of being locked out. When tagout is used, the following conditions will be met:

   (a) The tagout device will be attached at the same location the lockout device would have been attached, and

   (b) The tagout procedure will provide protection equal to that provided by a lockout procedure through the implementation of additional measures to supplement tags and protect workers.

   Examples of such measures are:

   (i) removing of an isolating circuit element (e.g., fuse),

   (ii) retraction of a draw-out circuit breaker from a switchboard (i.e., racking out the breaker)

   (iii) blocking a controlling switch (placement of a blocking mechanism over the operating handle of a disconnecting means so that the handle is blocked from being placed in the closed position).

   (iv) opening an extra disconnecting device (e.g., opening of a switch other than the disconnecting means which also opens the circuit between the source of power and the exposed parts on which work is to be performed; or opening of a switch for a control circuit that operates a disconnect that is itself open and disconnected from the control circuit or is otherwise disabled).

   (v) grounding the circuit upon which work is to be performed.

   (vi) removing a valve handle to reduce the likelihood of inadvertent energization, and/or

   (vii) positioning standby personnel at the tagout location.

5. Only authorized employees are permitted to implement a lockout/tagout procedure. Authorized employees will be trained in accordance with this policy and the OSHA Lockout/ Tagout standard. The names of those authorized to implement lockout/ tagout on a machine or piece of equipment will be identified in writing (e.g., on the equipment-specific energy control procedure form). The area Supervisor is responsible for identifying these individuals.
6. Lockout and tagout devices will be standardized for each facility. Authorized employees on all shifts will have access to these lockout/tagout devices.
7. All locks will be individually keyed (i.e., the "one lock – one key" rule), and will identify the individual to whom it is assigned. Identification will be accomplished by means of a collar and engraved tag attached to the lock, a durable label, or by using a tag that identifies the employee with the lock. The key to each lockout device will be in the sole possession of the employee to which it was assigned.

   Exception: Supervisors may have duplicate keys (or a master key) for the locks assigned to authorized employees under their supervision. Where duplicate master keys are used, the Supervisor will restrict and control access to the duplicate or master keys by keeping them in a locked key cabinet or other locked storage to which unauthorized employees do not have access.
8. Each lockout/tagout device will be removed from each energy isolating device by the employee who applied the lock or tag.

   Exception: When the authorized employee who applied the lockout or tagout device is not available to remove it, only the Supervisor, accompanied by another authorized, trained employee, may remove that device following the procedures set forth in section 5.10 of this policy.
9. If more than one employee is working on the equipment, multiple (individual) locks and/or tags will be used, or a group lockout/tagout procedure which ensures individual protection will be followed. (Refer to Specific Procedures for Group Lockout/ Tagout.)
10. For shift or personnel changes during a lockout or tagout, continuous protection will be maintained, and there will be an orderly transfer of lockout or tagout devices between the first (off-going) employee(s) and the relief employee(s).
11. Before performing any service or maintenance on cord and plug connected electrical equipment, the cord will be unplugged and the plug maintained under the exclusive control of the employee performing the servicing and/or maintenance. The plug will be considered to be under the exclusive control of the employee if it is physically in the possession of the employee, or within arms reach and in line of sight of the employee, or if the employee has affixed a personal lockout/tagout device on the plug.
12. Each disconnecting means for motors and appliances will be correctly and legibly marked to indicate its purpose, unless located and arranged so the purpose is evident. Each service, feeder, and branch circuit, at its disconnecting means or overcurrent device, will be correctly and legibly marked to indicate its purpose, unless located and arranged so the purpose is evident.

13. Employees will not be permitted to work in an area where they are likely to encounter electrical hazards unless they have been trained to recognize and avoid the electrical hazards to which they will be exposed. For the purposes of this policy, an electrical hazard is a recognizably dangerous electrical condition. Electrical conditions such as exposed energized parts and unguarded electric equipment that may become energized unexpectedly are examples of electrical hazards.
14. Only qualified persons will be allowed to work on or near exposed, energized electrical parts.
15. Where applicable, piping will be blanked and/or valves will be closed, chained, and locked and tagged. Piping systems will also be drained and decontaminated when necessary.
16. Piping systems should be identified and marked in accordance with ASME A13.1-1996—Scheme for the Identification of Piping Systems. Markings should identify the contents of the piping system, indicate direction of flow, and provide sufficient additional details such as temperature, pressure, etc. as necessary to identify the hazard. Identification may be accomplished by stenciling, the use of tape, or markers.
17. Whenever major replacement, repair, renovation or modification of machines or equipment is performed, the energy isolating device will be designed to accept a lockout device.
18. No new machines or equipment which utilize hazardous energy hazard will be purchased or installed unless appropriate energy isolating device(s) that are capable of being locked out are provided.

5.2. Preparation for Lockout/Tagout

1. All personnel affected by the intended service or maintenance task to be performed under lockout/tagout will be notified by the Supervisor or authorized employee before work is commenced.
2. Before commencing work, the authorized employee will determine the type(s), magnitude and hazards of energy to be controlled, and the method(s) or means to control the energy. The authorized employee will locate and identify all energy isolating devices that will be locked or tagged out.

5.3. Application of Lockout/Tagout (Sequence of Lockout or Tagout Procedure)

1. Verbally notify employees who work in the area that a lockout/tagout procedure is going to be used.
2. Shut the machine or equipment down by the normal stopping procedure.
3. Operate all energy isolating devices so that the equipment is isolated from all energy sources.
4. Dissipate or restrain all stored energy. Methods of dissipating or restraining stored energy include but are not limited to repositioning, blocking, bleeding down, etc. Stored electric energy which might endanger personnel will be released. Capacitors will be discharged and high capacitance elements will be short-circuited and grounded if the stored energy might endanger personnel.

5. Lockout and/or tagout the energy isolating device(s) with assigned individual lock(s) and/or tag(s) according to the method(s) selected.
6. Verify that the MEP has been effectively isolated and deenergization of the MEP has been accomplished.

5.4. Verification of Isolation

1. Prior to starting work on MEP that have been locked out or tagged out, the authorized employee will verify that isolation and deenergization of the MEP have been accomplished. This step of the procedure may involve a deliberate attempt to start the equipment which has been isolated, the use of appropriate test instruments, or visual inspection. A combination of these verification methods may be needed; the appropriate method or combination of methods will depend upon the type of machinery or equipment involved, the complexity of the system, and other factors.

   For example, after ensuring that no one is exposed, a qualified person may operate the equipment operating controls to verify that the equipment cannot be restarted.

   Note: After attempting to start the equipment, ensure that operating control(s) are returned to a safe condition (e.g., "neutral" or "off") before proceeding with service or maintenance. Another means of verifying is testing the machine or equipment with the appropriate test instruments. This method would be appropriate for use in cases involving electrical circuits and equipment, for example, where verification of isolation could be accomplished by a qualified person using a voltmeter to determine that there is no electrical energy available to the machine. Similar test equipment can be utilized to test for the presence of other energy types and sources. Visual inspection can be used to verify that isolation has been accomplished by confirming that all motion has stopped and that all coasting parts such as flywheels, grinding wheels, saw blades, etc., have come to rest. Visual inspection can also determine if switches, valves, breakers, etc. have been properly moved to and secured in the "off" or "safe" position.
2. Verification of energy isolation shall be monitored as frequently as necessary if there is a possibility of reaccumulation of stored energy

5.5. Release from Lockout/Tagout (Restoring the Machine or Equipment to Normal Operation)

1. After the servicing and/or maintenance is complete and the equipment is ready to be reenergized and/or made ready for normal production operations, a qualified person will verify the following so that the MEP can be safely energized:
   (a) All employees are in the clear;
   (b) The equipment is operationally intact;
   (c) Non-essential items (e.g., tools, jumpers, shorts, grounds and other such devices) have been removed; and
   (d) All guards, safety devices, etc. have been reinstalled.
2. Before the removal of locks and/or tags and the restoration of energy, employees exposed to hazards associated with re-energizing the MEP (i.e., affected employees or

persons) will be warned to stay clear of circuits and equipment and notified that the lockout/tagout devices will be removed.

3. Remove lockout/tagout devices.
4. Operate the energy isolating devices to restore energy.

5.6. Lockout/Tagout Interruption (Energized Testing, Positioning)

In situations where the energy isolating device(s) is locked/tagged and there is a need for energized testing or positioning of the MEP, the following additional sequence will apply:

1. Clear MEP of tools and materials;
2. Clear personnel;
3. Clear the energy isolating device(s) of locks/tags according to established procedure;
4. Proceed with testing, positioning;
5. De-energize and re-lock/tag energy isolating device(s) to continue the work; and
6. Verify isolation and deenergization of the MEP by use of appropriate meaans as described in the verification section of this policy.

5.7. Group Lockout/Tagout

1. Where more than one individual is required to lockout or tagout a MEP, a group lockout/tagout procedure will be followed. Where group lockout/tagout is used, primary responsibility for the safety of a set number of employees working under the protection of a group lockout or tagout device will be vested in one authorized employee (e.g., a group lockout/tagout "Coordinator"). The group lockout/tagout Coordinator is responsible for: overall lockout/tagout control; coordinating the affected workforces; and ensuring the continuity of protection during the work period. Specific procedural steps identified for the particular lockout or tagout will be followed.
2. The following general rules will be followed in group lockout/tagout situations:
   (a) Where applicable, each (authorized) employee will place his/her personal lockout/tagout device on the energy isolating device(s).
   (b) When an energy isolating device cannot accept multiple locks or tags, a multiple lockout or tagout device (e.g., hasp) will be used.
   (c) Each authorized employee will remove his/her own device(s) when he/she stops working on the equipment.
   (d) In certain cases, (i.e., where prescribed by a specific energy control procedure), a single lock may be used to lock out the MEP to minimize the number of personnel exposed to potentially hazardous equipment, with the key being placed in a lockout box or cabinet which allows the use of multiple locks to secure it. With the box or cabinet in a designated safe area, each (authorized) employee may then attach his/her own lock to secure the box or cabinet. As each person no longer needs to maintain his/her lockout protection, that person will remove his/her lock from the box or cabinet.

5.8. Shift or Personnel Changes

1. Specific procedures will be developed and used during shift or personnel changes to ensure the continuity and orderly transfer of lockout/tagout protection. In this regard, a "handshake" approach for transfer of lockout/tagout protection may be used. Under this method, off-going personnel will not leave the job site until the on-coming personnel attach their locks and tags to the appropriate isolating device(s). Verbal and/or written information will also be exchanged relative to the work in progress and the equipment's energy control status. The authorized employee with primary responsibility for the group (i.e., the Coordinator) will also be informed regarding the reason for and individuals involved with the switch.
2. For complex or group lockout/tagout applications, a "log and lock" approach may be used to ensure the orderly transfer and continuity of lockout/tagout protection. Under this method, a group lockout/tagout Coordinator is assigned overall responsibility for the area/operation, and is responsible for ensuring that a zero energy state is maintained during the work period and transition times. Under this method, a "control" lock belonging to the ABC Company authorized "Coordinator" will be applied to the MEP's energy isolating device(s) first, remain in place for the duration of the job, and will be the last removed from the energy isolating device(s). Each (authorized) employee will then place his/her personal locks/tags on the energy isolating device(s), hasp, lockout box or cabinet as appropriate. As each person no longer needs to maintain his/her lockout protection, that person will remove his/her lock from the energy isolating device, hasp, box or cabinet. For the purpose of personnel accounting, involved persons also log in and out on a master job control register to be kept by the Coordinator or designee.

5.9. Contractors and Outside Personnel

1. Whenever non-ABC Company servicing personnel are engaged in lockout or tagout activities, ABC Company personnel and the outside employer's personnel will inform each other of their respective lockout/tagout procedures.
2. When contractor employees perform service or maintenance activities on ABC Company equipment which requires lockout/tagout, application of a "control" lock by a qualified and authorized ABC Company employee may be required. Where applicable, the ABC Company control lock will be applied first, will remain in place for the duration of the job, and will be the last removed from the energy isolating device(s).

5.10. Emergency Removal of LOTO Devices

During emergencies or other times when the authorized employee who applied a lockout or tagout device is not available to remove it (or cannot be identified), and there is an urgent need to remove the device, that device may be removed by his or her Supervisor provided that the following steps are performed:

1. The Supervisor must make a determined effort to locate the authorized employee and must verify that the he/she is not at the facility. In the case of an abandoned or unidentifiable lock, all reasonable efforts will be made to identify the lock/tag owner.

2. The Supervisor must make every reasonable effort to contact the authorized employee who placed the device. This should include a telephone call to the employee's home or other location.
3. If the employee is contacted, the Supervisor must inform the employee that his or her device(s) are being removed.
4. When removing the device(s), the Supervisor must be accompanied by another authorized employee.
5. The Supervisor and accompanying authorized employee must verify that it is safe to remove the device(s).
6. The Supervisor may then use a master or duplicate key to remove the lockout/tagout device(s), or the lock may be cut off if a key is not available.
7. The Supervisor will ensure that the authorized employee is informed about the removal of the device before he/she resumes work at the facility. The authorized employee will be presented with the removed lock, and informed of the reasons for the emergency removal upon his/her return to work.

6.0. ENERGY CONTROL PROCEDURES (ECP)

1. When lockout/tagout is required, specific written energy control procedures will be developed and used for each machine, equipment or process.
2. ECPs should include:
   (a) A specific statement of the intended use of the procedure;
   (b) The type(s) and magnitude(s) of hazardous energy associated with the MEP;
   (c) The type(s) and location(s) of the isolating device(s);
   (d) Involved personnel;
   (e) Specific procedural steps for shutting down, isolating, blocking, and securing machines or equipment to control hazardous energy;
   (f) Specific procedural steps for the placement, removal, and transfer of Lockout/Tagout devices and the responsibility for them;
   (g) Specific requirements for testing a machine or equipment to determine and verify the effectiveness of lockout devices, and other energy control measures
   (h) The specific steps for reenergizing the machine and restoring it back to normal operation.

   Note: The attached form or equivalent may be used to document the ECPs.
3. Authorized employees will follow the specific energy control procedure when performing lockout/tagout on a piece of equipment.
4. The specific procedures will be modified as necessary to reflect changes in the equipment, the method or location of isolation, the energy sources, and/or authorized employees.

7.0. TRAINING

1. Training will be provided to employees to ensure that they understand the purpose and function of the energy control program and have the knowledge and skills required for the safe application, use and removal of energy controls.
2. Each authorized employee will receive training in the recognition of applicable hazardous energy sources, the type(s) and magnitude of the energy available in the workplace, the methods and means necessary for energy isolation and control, and the means of verification of control.
3. Each affected employee will be instructed in the purpose and use of the Lockout/Tagout policy and procedure, and the prohibition relating to attempts to restart or reenergize MEP which are locked out or tagged out.
4. When tagout systems are used, employees will be trained in the limitations of tags pursuant to 29 CFR 1910.147 (c)(7)(C)(ii).
5. Retraining will be provided whenever there is a change in job assignment, a change in machines, equipment or processes that present a new hazard, or when there is a change in the energy control procedures or revision of control methods. Additional retraining will also be conducted whenever a periodic audit or Supervisory observation of the energy control procedures indicates deviation from or inadequacies in an employee's knowledge or use of the energy control procedures. The retraining will reestablish employee proficiency and introduce new or revised control methods and procedures, as necessary.
6. Documentation of all employee training will be kept to certify that training has been accomplished and is up-to-date. Documentation will include each employee's name, signature, employee number, training location, work location, instructor, date(s) of training, and duration of training.

8.0. PERIODIC INSPECTIONS

1. An inspection of each energy control procedure will be conducted at least annually by the area Supervisor or designee. The person performing the inspection will be an authorized employee other than the one(s) using the energy control procedure being inspected.
2. The inspection will include a review between the inspector and each authorized employee (and affected employee if a tagout system alone is used) of the employee's responsibilities under the energy control procedure being inspected. The inspection will determine the following:
   - whether the steps in the energy control procedure are being followed;
   - whether the employees involved know their responsibilities under the procedure; and
   - whether the procedure is adequate to provide the necessary protection, and what changes, if any, are needed.
3. The Supervisor will ensure that appropriate action is taken to ensure that any deficiencies identified during the inspection are corrected.

4. The Supervisor will certify that the periodic inspections have been performed. The certification will identify the machine or equipment on which the energy control procedure was being utilized, the date of the inspection, the employees included in the inspection, and the person performing the inspection. The attached *Lockout/Tagout Periodic Inspection Form* may be used for this purpose.

9.0. PROTECTIVE MATERIALS AND HARDWARE

1. Locks, tags, chains, wedges, key blocks, adapter pins, self locking fasteners, or other hardware will be provided as necessary for isolating, securing or blocking of machines or equipment from energy sources.
2. Lockout devices will be singularly identified; will be the only device(s) used for controlling energy; will not be used for other purposes; and will meet the following requirements:
   (a) Durable
      (i) Lockout and tagout devices will be capable of withstanding the environment to which they are exposed for the maximum amount of time that exposure is expected.
      (ii) Tagout devices will be constructed and printed so that exposure to weather conditions or wet and damp locations will not cause the tag to deteriorate or the message on the tag to become illegible.
      (iii) Tags will not deteriorate when used in corrosive environments such as areas where acid or alkali chemicals are handled and stored.

   (b) Standardized.

      Lockout and tagout devices will be standardized within the facility in at least one of the following criteria: color; shape; or size; and additionally, in the case of tagout devices, print and format will be standardized.

   (c) Substantial
      (i) Lockout and tagout devices will be substantial enough to prevent removal without the use of excessive force or unusual techniques, such as with the use of bolt cutters or other metal cutting tools.
      (ii) Tagout devices, including their means of attachment, will be substantial enough to prevent accidental or inadvertent removal. Tagout device attachment means will be of non-reusable type, attachable by hand, self-locking, and non-releasable with a minimum unlocking strength of no less than 50 pounds and having the general design and basic characteristics of being at least equivalent to a one-piece all-environment-tolerant nylon cable tie.

   (d) Identifiable. Lockout and tagout devices will indicate the identity of the employee applying the device(s).
3. Tagout devices will warn against hazardous conditions if the machine or equipment is energized and will include a legend such as the following: Do Not Start, Do Not Open, Do Not Close, Do Not Energize, Do Not Operate.

<table>
<tr><th colspan="5">ABC Company
ENERGY CONTROL PROCEDURE</th></tr>
<tr><td colspan="2">Equipment Description:</td><td colspan="3"></td></tr>
<tr><td colspan="2">Equipment Number:</td><td colspan="3"></td></tr>
<tr><td colspan="2">Job(s):</td><td colspan="3"></td></tr>
<tr><th colspan="2">Hazardous Energy</th><th colspan="3">Isolating Devices</th></tr>
<tr><td>Type</td><td>Magnitude</td><td>Type</td><td>Location</td><td>ID Number</td></tr>
<tr><td></td><td></td><td></td><td></td><td></td></tr>
<tr><td></td><td></td><td></td><td></td><td></td></tr>
<tr><td></td><td></td><td></td><td></td><td></td></tr>
</table>

<table>
<tr><th colspan="2">Involved Personnel</th></tr>
<tr><td>Authorized Employees</td><td>Affected Employees</td></tr>
<tr><td></td><td></td></tr>
</table>

✔	**Step #**	**Procedure**
	1	
	2	
	3	
	4	
	5	use additional sheets if necessary

Preparer's name: Date:

Lockout/Tagout
Periodic Inspection Certification

Facility/Site:	
Department:	
Name/Identification of Machine, Equipment, or Process:	
Energy Control Procedure Being Reviewed:	
Date:	

Name of Employee(s) Being Reviewed (use additional sheets if necessary)

1.	6.	11.
2.	7.	12.
3.	8.	13.
4.	9.	14.
5.	10.	15.

Inspection items–Review the Energy Control Procedure and employee responsibilities with the involved employees and complete the following:	**Yes**	**No**
1. Are the steps in the energy control procedure being properly followed? (If no, provide a detailed description of the problem below, along with a description of any corrective action taken or planned.)		
2. Do the involved employees understand their responsibilities under the procedure? (If no, provide a detailed description of the problem and any corrective action needed below.)		
3. Are there any inadequacies in any employee's knowledge, abilities, or use of the procedures? (If yes, provide a detailed description of the problem and any corrective action needed below.)		
4. Is the procedure adequate to provide the necessary protection? (If no, provide a detailed description of the problem and any corrective action needed below.)		

Corrective Action–Use the space provided below to describe any problems identified during the inspection, along with a description of any corrective action needed. Appropriate action must be taken to ensure that the deficiencies are corrected. This may involve making changes to the procedure, providing retraining to employees, and/or taking other steps needed to ensure compliance.

Person(s) Conducting the Inspection

Name (print):	**Signature:**	**Title/Department:**

Glossary

Ampere

The unit of measurement for electric current. It represents the rate at which current flows through a resistance of one ohm by a pressure of one volt.

Attachment plug

A device that establishes a connection between the conductors in its flexible cord and those in the receptacle into which it is inserted.

Affected employee

An employee whose job requires him or her to use a machine on which maintenance is being performed under lockout or tagout, or whose job requires him or her to work in an area in which such maintenance is being performed. An affected employee becomes an authorized employee when that employee's duties include performing servicing or maintenance. (*see* Servicing/maintenance, Authorized employee)

Authorized employee

A person who locks out or tags out machines in order to perform maintenance on them.

Bonding jumper

A reliable conductor used to ensure the required electrical conductivity between metal parts that are required to be electrically connected.

Branch circuit

The portion of a wiring system that extends beyond the final overcurrent device protecting the circuit.

Capable of being locked out

An energy-isolating device is capable of being locked out if it has a hasp or other means of attachment to which a lock can be affixed or it has a locking mechanism built into it. Other energy-isolating devices are capable of being locked out if lockout can be achieved without needing to dismantle, rebuild, or replace them or permanently alter their energy control capability.

Circuit breaker

A device designed to open and close a circuit manually and to open the circuit automatically when overloaded without injury to itself (if it is properly applied within its rating).

Controller

A device or group of devices that serves to govern in some predetermined manner the electric power delivered to the apparatus to which it is connected.

Current

The flow of electricity in a circuit. It is expressed in amperes and represents an amount of electricity.

Disconnecting means

A device, group of devices, or other means by which the conductors of a circuit can be disconnected from their source of supply.

Electrical energy

Energy derived from generated electrical power, sources of static electricity, or electrical storage devices such as batteries and capacitors.

Energized

Connected to an energy source or containing residual or stored energy.

Energy-isolating device

A mechanical device that physically prevents the transmission or release of energy, including but not limited to the following:

- Manually operated electrical circuit breakers.
- Disconnect switches.
- Manually operated switches by which the conductors of a circuit can be disconnected from all ungrounded supply conductors, and, in addition, no pole can be operated independently.
- Line valves.
- Blocks.
- Any similar devices used to block or isolate energy.

Push buttons, selector switches, and other control circuit type devices are not energy isolating devices.

Energy source

Any source of electrical, mechanical, hydraulic, pneumatic, chemical, thermal, or other energy.

Ground

A conducting connection, intentional or accidental, between an electrical circuit or piece of equipment and earth or some other conducting body.

Grounded

Connected to earth or some other conducting body.

Hot tap

A procedure used in the maintenance of equipment under pressure, such as pipelines, vessels, or tanks, which involves welding to install connections or appurtenances. It is commonly used to replace or add sections of pipeline without the interruption of service for air, gas, water, steam, and petrochemical distribution systems.

Job-lock

A device placed on a lockbox to ensure the continuity of energy isolation during a multi-shift operation. A key to the job-lock is controlled by each assigned primary authorized employee from each shift.

Job-tag with tab

A special tag for the tagout of energy isolating devices during group lockout/tagout procedures. The tab of the tag is removed for insertion into the lockbox. The lockout/tagout procedure should require that the tagout job-tag cannot be removed until the tab is rejoined to it.

Lockout

The placement of a lockout device on an energy isolating device in accordance with an established procedure, thus ensuring that the energy isolating device and the equipment being controlled cannot be operated until the lockout device is removed.

Lockout device

A device that uses a positive means such as a key or combination lock to hold an energy isolating device in the safe position and prevent the energizing of a machine. Included are blank flanges and bolted slip blinds.

Master lockbox

The lockbox into which all keys and tabs from the lockout or tagout devices securing a machine are inserted and which is secured by a job-lock during multi-shift operations.

Master tag

A document used as an administrative control and accountability device. The master tag is normally controlled by operations department personnel. It is a personal tagout device if each employee personally signs on and signs off on it and if the tag clearly identifies each authorized employee who is being protected by it.

Mechanical energy

The kinetic energy associated with the moving parts of mechanical systems. Mechanical energy can be manifested in linear or rotational mechanical motion.

Normal production operations

The use of a machine to perform its intended production function.

Ohm

The unit of measurement of electrical resistance; it represents the amount of resistance that permits current to flow at the rate of one ampere under a pressure of one volt.

Potential, electrical

The difference in voltage between two points of a circuit.

Potential energy

Stored energy that can be released as hazardous mechanical energy. Potential energy is associated with pressure in pneumatic, hydraulic, and vacuum systems, with springs under tension or compression, and with gravity (e.g., raised loads).

Primary authorized employee

The authorized employee who exercises overall responsibility for adherence to the company lockout/tagout procedure.

Principal authorized employee

An authorized employee who oversees or leads a group of servicing/maintenance workers.

Qualified person

A person familiar with the construction and operation of equipment and the hazards involved.

Servicing and maintenance

Workplace activities such as constructing, installing, setting up, adjusting, inspecting, modifying, and maintaining or servicing machines or equipment. These activities include lubricating, cleaning, or unjamming machines or equipment and making adjustments or tool changes, where the employee may be exposed to the unexpected energization or startup of the equipment or release of hazardous energy.

Satellite lockbox

A secondary lockbox to which each authorized employee affixes his or her personal lock or tag.

Setting up

Any work performed to prepare a machine or equipment to perform its normal production operation.

Tagout

The placement of a tagout device on an energy-isolating device in accordance with an established procedure to indicate that it and the equipment being controlled may not be operated until the tagout device is removed.

Tagout device

A prominent warning device, such as a tag and a means of attachment, which can be securely fastened to an energy isolating device in accordance with an established procedure to indicate that it and the equipment being controlled may not be operated until the tagout device is removed.

Thermal energy

The energy resulting from high or low temperature. Thermal energy can result from mechanical work, radiation, chemical reaction, or electrical resistance.

Utilization equipment

Equipment that uses electric energy for mechanical, chemical, heating, lighting, or similar useful purposes.

Volt

The unit of measurement of electromotive force. One volt sends a current of one ampere through a resistance of one ohm.

Voltage

The force or pressure that causes electric current to flow in an electric circuit. Its unit of measurement is the volt.

Work permit

A control document that authorizes specific tasks and procedures to be accomplished.

Index